AF352107

**Normal Tissue Reactions in Radiotherapy and Oncology**

Frontiers of Radiation
Therapy and Oncology

Vol. 37

Series Editors

*John L. Meyer*  San Francisco, Calif.
*W. Hinkelbein*  Berlin

KARGER

International Symposium, Marburg, Germany, April 14–16, 2000

# Normal Tissue Reactions in Radiotherapy and Oncology

Volume Editors

*Wolfgang Dörr* Dresden
*Rita Engenhart-Cabillic* Marburg
*Jörg S. Zimmermann* Munich

39 figures, and 35 tables, 2002

Basel · Freiburg · Paris · London · New York ·
New Delhi · Bangkok · Singapore · Tokyo · Sydney

# Frontiers of Radiation Therapy and Oncology

Founded 1968 by J.M. Vaeth, San Francisco, Calif.

**Prof. Dr. W. Dörr**
Department of Radiotherapy and Radiation Oncology
Medical Faculty Carl Gustav Carus
Technical University, Dresden, Germany

**Prof. Dr. R. Engenhart-Cabillic**
Department of Radiotherapy and Radiation Oncology
Medical Centre of Radiology
University Hospital, Philipps-University, Marburg, Germany

**Dr. Jörg S. Zimmermann**
Department of Brachytherapy
Munich Radiation Cancer Center (MRCC)
Munich, Germany

Library of Congress Cataloging-in-Publication Data

Normal tissue reactions in radiotherapy and oncology: international symposium,
Marburg, April 14–16, 2000 / volume editors, Wolfgang Dörr, Rita Engenhart-Cabillic,
Jörg S. Zimmermann.
     p. ; cm. – (Frontiers of radiation therapy and oncology; vol. 37)
    Includes bibliographical references and index.
    ISBN 3805572840 (hard cover : alk. paper)
    1. Radiotherapy–Side effects–Congresses. I. Dörr, Wolfgang, Prof. Dr. II.
Engenhart-Cabillic, Rita, 1953- III. Zimmermann, Jörg S. IV. Series.
    [DNLM: 1. Radiotherapy–adverse effects–Congresses. 2.
Neoplasms–radiotherapy–Congresses. 3. Radiation Effects–Congresses. QZ 269 N842 2002]
    RM849.N67 2002
    615.8'42–dc21                          2001038675

Bibliographic Indices. This publication is listed in bibliographic services, including Current Contents® and Index Medicus.

Drug Dosage. The authors and the publisher have exerted every effort to ensure that drug selection and dosage set forth in this text are in accord with current recommendations and practice at the time of publication. However, in view of ongoing research, changes in government regulations, and the constant flow of information relating to drug therapy and drug reactions, the reader is urged to check the package insert for each drug for any change in indications and dosage and for added warnings and precautions. This is particularly important when the recommended agent is a new and/or infrequently employed drug.



© Copyright 2002 by S. Karger AG, P.O. Box, CH–4009 Basel (Switzerland)
www.karger.com
Printed in Switzerland on acid-free paper by Reinhardt Druck, Basel
ISSN 0071–9676
ISBN 3–8055–7284–0

# Contents

Outcome, Forensic Aspects and Rehabilitation



Clinical Management I



Radio- and Cytoprotection



Clinical Management II

# Preface

The International Symposium on Normal Tissue Reactions in Radiotherapy and Oncology was held in Marburg, Germany, from April 14 to 16, 2000. In this issue of *Frontiers of Radiation Therapy and Oncology* papers that were presented at this conference are published.

Since the discovery of X-rays and radioactivity more than 100 years ago, radiation oncology underwent extremely dynamic changes in treatment planning and establishment of irradiation techniques for tumor therapy. Some decades ago, deep X-ray therapy was the dominating treatment approach. This was associated with the use of high radiation doses at the beam entrance site, i.e. in skin and subcutaneous soft tissues. Therefore, acute skin reactions to radiotherapy were dose-limiting, and led to dose adjustment (skin erythema dose). In contrast, with linear accelerators and modern computer-aided treatment planning techniques, nearly any dose distribution within the body can be achieved. Hence, superficial, visible radiation effects, e.g. in skin or oral mucosa, are no longer valid indicators for side effects in other organs or tissues.

The physical parameters of radiation treatment have been optimized for adjusting the dose to the tumor tissue by 3-dimensional treatment planning, multiple field treatments in 3-dimensional conformal radiotherapy [e.g. 1], intensity-modulated radiotherapy [e.g. 2, 3] or radiosurgery [4]. For further optimization of the ratio between tumor control and side effects, biological parameters must be included in treatment planning. This has led to a variety of mathematical models of normal tissue complication probability, which include biological tissue organization, irradiated tissue volume and other parameters. The problem with these modelling exercises is the quality of data which they are based on. Insufficient information can readily result in misleading conclusions. Therefore, permanent improvement of the data base and adjustment of the model parameters are required to improve model predictions [5].

The development in radiooncological treatment protocols was accompanied by substantial changes in the pattern of side effects. Acute skin reactions today are of minor importance, while radiation sequelae in other organs and tissues play the predominant role. This particularly refers to late side effects, which are assumed to be irreversible and often substantially impair quality of life. However, acute side effects of treatment not only result in temporary effects on the patients. They may also necessitate hospitalization or treatment interruptions, with significant economic and biological consequences. Moreover, acute radiation reactions can affect late treatment effects by a consequential component in some tissues, such as gut, urinary tract or oral mucosa [6]. Therefore, therapeutic modulation of acute as well as of late effects will be of benefit to the patients.

Simultaneous administration of cytostatic drugs in multimodal treatment regimens has introduced further changes in the spectrum of side effects of tumor therapy, particularly with regard to acute reactions; examples are presented in this issue [7–9]. Amplification of side effects of radiotherapy can occur if the target cells of both agents are identical. Also, combinations of side effects may be observed, which are different from those after either radiation or drug alone, if different target cells or tissues are affected. In combination with surgery, effects of radiation and/or chemotherapy on preirritated or predamaged tissues, such as blood vessels, lymph drainage, and on wounds must be considered [10, 11].

During the last century, the outcome of tumor therapy has progressively improved. Severe, life-threatening complications, both at acute and intermediate time points, have become rarer events. However, increased tumor cure and survival rates and prolonged survival times in consequence allow for the manifestation of less severe changes, and also of late changes with latent times of 10 years and more. Prolonged survival does also allow for the manifestation of secondary, radiation-induced tumors with latent times in the range of 10–20 years [12]. Therefore, a sufficiently long follow-up by radiation oncologists is necessary in order to identify the full spectrum of side effects, to quantitate the precise incidence, and to define the consequences on the patients' quality of life [e.g. 13]. In pediatric radiotherapy, the follow-up must focus on organs at risk which are partially different from those in adults [14].

Ideal curative radiation therapy implies that the maximum dose given to the tumor, aiming at local tumor control, must be accompanied by minimum doses to normal tissues. However, normal tissues, i.e. normal structures both within the tumor and in the margin around the tumor, must necessarily be included in the maximum dose volume (planning target volume). Moreover, normal tissues in the beam entrance and the exit channel may be exposed to lower but still significant radiation doses.

Therefore, documentation, quantification and publication of side effects associated with a specific treatment protocol, not only with radiation alone but also in multimodal regimens, are a major prerequisite for quality control in tumor therapy. Moreover, detailed knowledge of side effects is the basis for suitable information of the patients, which has recently gained increasing importance in the face of personal liability of the physicians involved in the treatment and the liability of the institutions where the treatment is performed [15].

Development of effective approaches to the prophylactic and therapeutic management of side effects is one of the major tasks of modern (radiation) oncology. The view of radiobiology of normal tissue effects of radiation exposure has changed during the last decade [16]. Classical cellular radiobiology assumed that acute effects are exclusively due to sterilization of (hypothetical) stem cells. Hence, modulation of stem cell survival is restricted to the time when the radiation damage occurs, and would be limited to treatment parameters such as fractionation or overall treatment time [17], or simultaneous treatment, e.g. with radioprotective agents [18].

Recently, an increasing pool of experimental results indicates postirradiation processing of radiation damage, which is associated with substantial changes in tissue protein synthesis and release. Based on detailed knowledge of these pathogenetic mechanisms eventually resulting in the impairment of tissue and organ function, radiation responses may well be modulated after the radiation insult [16, 19].

General side effects of tumor therapy include fatigue associated with anemia, which may be managed by the correction of blood hemoglobin levels, e.g. by erythropoietin [20]. After successful tumor therapy, but also during treatment, effective rehabilitation programs focussing on both physical activity as well as psychosocial treatment effects can significantly improve the patients' quality of life [21, 22].

Clinical management approaches to dealing with various side effects of radio- and radiochemotherapy must be designed for the individual organs affected. Some examples have been presented at the conference [14, 23]. They may either be symptomatic or based on the pathogenetic principles underlying the radiation reaction. For this, proper preclinical normal tissue studies, including animal models, must be exploited. In order to guarantee the selectivity of the management options tested, additional studies with suitable in vitro and – even more importantly – in vivo tumor models, including human xenografts grown in nude mice, are essential.

A major prerequisite for the control of beneficial effects of such management is detailed scoring and documentation of side effects, if necessary supplemented with additional diagnostic procedures, such as ultrasound for skin changes [24], PET for lung changes [25] or laboratory analyses e.g. for pancreatic changes [26]. These data may then be used as a platform to design novel strategies for the prophylactic or therapeutic management of side effects.

Clinical studies, prospective, randomized and with a sufficient number of patients, are then required to translate the experimental findings into clinical practice. Bringing together medical, biological, molecular and physical expertise, research in the field of radiation-induced side effects represents a true interdisciplinary effort.

*Wolfgang Dörr*, Dresden
*Rita Engenhart-Cabillic*, Marburg
*Jörg S. Zimmermann*, Munich

## References


1   Nutting CM, Bedford JL, Cosgrove VP, Tait DM, Deamaley DP, Webb S: Intensity-modulated radiotherapy reduces lung irradiation in patients with carcinoma of the oesophagus. Front Radiat Ther Oncol. Basel, Karger, 2001, vol 37, pp 128–131.
2   De Neve W, Claus F, Duthoy W, de Meerleer G, de Wagter C: Intensity modulation techniques for improvement of normal tissue tolerance. Front Radiat Ther Oncol. Basel, Karger, 2001, vol 37, pp 163–173.
3   Nutting CM, Deamaley DP: Prostate cancer – The Royal Marsden conformal experience. Front Radiat Ther Oncol. Basel, Karger, 2001, vol 37, pp 196–199.
4   Gross MW, Engenhart-Cabillic R: Normal tissue reactions after linac-based radiosurgery and stereotactic radiotherapy. Front Radiat Ther Oncol. Basel, Karger, 2001, vol 37, pp 140–150.
5   Burman CM: Fitting of tissue tolerance data to analytic function: Improving the therapeutic ratio. Front Radiat Ther Oncol. Basel, Karger, 2001, vol 37, pp 151–162.
6   Dörr W, Hendry JH: Consequential late effects in normal tissues. Radiother Oncol, in press.
7   Dunst J: Normal tissue reactions during and after radiochemotherapy. Front Radiat Ther Oncol. Basel, Karger, 2001, vol 37, pp 69–77.
8   Hartmann JT: Long-term effects of platin and anthracycline derivatives and possible prevention strategies. Front Radiat Ther Oncol. Basel, Karger, 2001, vol 37, pp 92–100.
9   Wilkowski R, Heinemann V, Stoffregen C: Gemcitabine (Gemzar®) and radiotherapy – Is it feasible? Front Radiat Ther Oncol. Basel, Karger, 2001, vol 37, pp 78–83.
10  Lippert BM, Dünne AA, Werner JA: Side effects in multimodal treatment concepts in carcinomas of the head and neck. Front Radiat Ther Oncol. Basel, Karger, 2001, vol 37, pp 185–190.
11  Von Knobloch R, Wille S, Hofmann R: Clinical side effects after radical prostatectomy. Front Radiat Ther Oncol. Basel, Karger, 2001, vol 37, pp 191–195.
12  Kodym R: Secondary malignancies after multimodal treatment regimens. Front Radiat Ther Oncol. Basel, Karger, 2001, vol 37, pp 84–91.
13  Vordermark D, Sailer M, Flentje M, Thiede A, Kölbl O: Impaired sphincter function and good quality of life in anal carcinoma patients after radiotherapy: A paradox? Front Radiat Ther Oncol. Basel, Karger, 2001, vol 37, pp 132–139.
14  Dieckmann K, Widder J, Pötter R: Long-term side effects of radiotherapy in survivors of childhood cancer. Front Radiat Ther Oncol. Basel, Karger, 2001, vol 37, pp 57–68.
15  Resch-Holeczke A, Ofner H, Pötter R: Personal and institutional liability. Front Radiat Ther Oncol. Basel, Karger, 2001, vol 37, pp 43–48.
16  Dörr W: Acute radiation effects in normal tissues – Translational aspects of biological research. Front Radiat Ther Oncol. Basel, Karger, 2001, vol 37, pp 1–8.

17    Maciejewski B, Miszczyk L, Tamawski R, Skladowski K: How the game is played – Challenge between therapeutic benefit and acute toxicity in fractionated radiotherapy. Front Radiat Ther Oncol. Basel, Karger, 2001, vol 37, pp 174–184.
18    Strnad V, Sauer R: Radioprotection of head and neck tissue by amifostine. Front Radiat Ther Oncol. Basel, Karger, 2001, vol 37, pp 101–111.
19    Trott K-R: Experimental radiotherapy of late responding tissues – Recent advances and future development. Front Radiat Ther Oncol. Basel, Karger, 2001, vol 37, pp 9–16.
20    Littlewood TJ: Efficacy and quality of life outcomes of epoietin alpha in a double-blind, placebo-controlled multicenter study of cancer patients receiving nonplatinum-containing chemotherapy. Front Radiat Ther Oncol. Basel, Karger, 2001, vol 37, pp 34–37.
21    Dimeo F: Radiotherapy-related fatigue and exercise for cancer patients: A review of the literature and suggestions for future research. Front Radiat Ther Oncol. Basel, Karger, 2001, vol 37, pp 49–56.
22    Strobel E-S: Physical and psychosocial rehabilitation of cancer patients. Front Radiat Ther Oncol. Basel, Karger, 2001, vol 37, pp 38–42.
23    Fraunholz IB, Schopohl B, Falk S, Boettcher HD: Cytohormonal status and acute radiation vaginitis. Front Radiat Ther Oncol. Basel, Karger, 2001, vol 37, pp 112–120.
24    Wratten C, Kilmurray J, Wright S, O'Brien P, Back M, Hamilton C, Denham J: A study of high frequency ultrasound to assess cutaneous oedema in conservatively managed breast. Front Radiat Ther Oncol. Basel, Karger, 2001, vol 37, pp 121–127.
25    Nestle U, Hellwig D, Fleckenstein J, Walter K, Ukena D, Rübe C, Kirsch C-M, Baumann M: Comparison of early pulmonary changes in 18FDG-PET and CT after combined radiochemotherapy for advanced non-small cell lung cancer: A study in 15 patients. Front Radiat Ther Oncol. Basel, Karger, 2001, vol 37, pp 26–33.
26    Horst E, Seidel M, Micke O, Rübe C, Schäfer U, Willich N: Functional evaluation of the human pancreas before and in the early period after hyperfractionated accelerated radiochemotherapy. Front Radiat Ther Oncol. Basel, Karger, 2001, vol 37, pp 17–25.

Dörr W, Engenhart-Cabillic R, Zimmermann JS (eds): Normal Tissue Reactions in Radiotherapy and Oncology. Front Radiat Ther Oncol. Basel, Karger, 2002, vol 37, pp 1–8

# Acute Radiation Effects in Normal Tissues – Translational Aspects of Biological Research

*Wolfgang Dörr*

Klinik und Poliklinik für Strahlentherapie und Radioonkologie, Medizinische Fakultät Carl Gustav Carus, Technische Universität Dresden, Deutschland

Acute side effects of cancer treatment by radiation, with or without chemotherapy, gain increasing relevance with the application of modern, aggressive treatment protocols. Severe acute treatment sequelae can be *dose-limiting*, and hence may affect tumor control. Moreover, in tissues such as the intestine, oral mucosa or urinary bladder, the *consequential component* of late treatment effects can become dominant if treatment protocols which aggravate acute effects are introduced. Modulation of acute radiation reactions must hence be considered a potent strategy to improve the therapeutic ratio in effective cancer treatment.

These clinical strategies, aiming at improvement of therapeutic outcome, may be established by a *classical translational research chain*. The basis for the modulation of acute normal tissue responses is laid by initial investigations in in vitro models which include molecular and cell biology approaches. Results from these investigations must be incorporated and further validated in in vitro research projects. Once the significance of the data for tissue responses to radiation injury is indicated, this has to be tested in radiobiological studies in relevant and suitable in vivo models. In parallel, all approaches for the modification of (acute) normal tissue reactions must be tested for their selectivity in studies with relevant tumor models. Eventually, clinical phase I/II, and phase III trials must be performed. This sequence represents forward translation of information [1]. The reverse sequence, with transfer of experience from clinical practice to preclinical research at all levels, can provide a new input into the chain. This has a stimulatory but also regulatory function for the initiation of new research projects which are of clinical relevance [1].

In the present paper the translational chain will be discussed with regard to acute radiation responses, with its potentials and pitfalls. Examples will be presented where successful transfer of information through the chain resulted in novel clinical approaches, or – which is also relevant – prevented inadequate clinical strategies. Examples for the flow of misleading information will be given as well.

## Pathogenesis of Acute Side Effects of Radiotherapy in Normal Tissues

The pathogenesis of acute radiation effects in normal tissues, from the view of classical cellular radiobiology, can be described by a number of well-defined steps: induction, progression and manifestation of tissue damage, followed by restoration [3, 13]. Tissue radiation injury is induced by sterilization of a significant number of cells of the relevant cell population, i.e. the *target cells*. For most acute radiation effects, this population is well known: clonogenic keratinocytes in skin and oral mucosa, gastrointestinal mucosal stem cells, or bone marrow stem cells. However, for epithelial cells, in contrast to bone marrow cells, no direct differentiation between stem and proliferating non-stem cells is possible – at present there is no known stem cell marker. For some other acute effects, like changes in urinary bladder storage function, the target cell population is still undefined.

Irradiation of proliferating cells in the tissue results in loss of the proliferative capacity, despite a limited number of residual, so-called 'abortive' mitoses. In contrast, physiological cell loss due to tissue function, e.g. by mechanical stress at epithelial surfaces, is unaffected by radiation and continues at its normal rate during and after irradiation. Due to the *hierarchical structure* of most of the acutely responding tissues, loss or impairment of cell production hence results in progressive tissue hypoplasia and eventually in complete loss of functional cells. This pathogenetic concept of acute effects in hierarchical tissues does not include further processes, like the acute vascular response, the relevance of which for the epithelial radiation effect is unknown at present. Based on surviving stem cells, both original cell numbers and the original radiosensitivity of the tissue are restored. In other tissues, such as the urinary bladder, more complex processes are observed, and changes in cell function in the urothelium seem to be the major determinant for the radiation effect [8, 33].

## Classical Radiobiological Research

Radiobiological studies of classical parameters of acute radiation responses, i.e. dose fractionation and repopulation [13, 28, 29, 34], were performed in in vitro systems as well as in experimental animals, and were supported by analyses of clinical data. These investigations yielded qualitative and quantitative knowledge of these factors of the radiation response for individual tissues, including those which display acute responses. In numerous experimental projects, changes in normal tissue radiation tolerance with the number or size of fractions (*capacity of recovery from sublethal damage*), but also with the time interval between fractions (*recovery kinetics*) were defined quantitatively [29]. It has turned out that the fractionation effect is one of the exceptional situations where direct quantitative transfer of data from animals to human tissues is possible [29, 30]. These studies yielded detailed results of values for the $\alpha/\beta$-ratio and half-times of recovery which – in combination with data derived from clinical studies – now allow calculation of isoeffective doses for novel fractionation protocols in clinical radiotherapy.

Another example, which is more specific for acute radiation sequelae, is the effect of overall treatment time. The biphasic time course of *repopulation*, with an initial latent period and a very effective subsequent regeneration response has been defined in a number of experimental systems, as summarized in Dörr [4]. Based on early studies in animals, the effect of changes in overall treatment time, e.g. in the continuous, hyperfractionated, accelerated radiotherapy (CHART) protocol [2], on the intensity of acute radiation effects can be anticipated and the doses and dose intensities can be adjusted to avoid intolerable acute reactions. The mechanisms underlying this response were defined as well in detailed radiobiological and cell kinetic studies [9, 10]. These mechanisms may be summarized as the 3 As of repopulation: *a*symmetry loss and an *a*ccelerated rate of stem cell divisions, along with *a*bortive divisions of sterilized cells [4].

Classical radiobiological studies, both in vitro and in vivo, revealed data for the estimation of the effectiveness of *unconventional radiation qualities*, including neutrons, pions, and, more recently, heavy ions [16]. These results were incorporated into models used for radiotherapy dose planning and dose distribution estimates in cancer treatment exploiting the specific biological features of these radiations [e.g. 26].

There are examples from the past which demonstrate that the translational research chain does not always function. A number of *radioprotective agents* have shown promising results in in vitro studies in cell cultures, and also in in vivo testing. The dose modification factors reported range from 1 to about 3 [16]. However, clinical application of these factors showed that side effects, like nausea and vomiting, diarrhea, hypotension, neurotoxicity and others, in most cases rendered this approach unfeasible. Moreover, the selectivity of this radioprotective treatment for normal tissues is questionable. In these examples, further and relevant preclinical experimentation would have been required to prevent the clinical approaches.

Another example is the exploitation of *cell cycle effects*, where the translational chain led to misleading results. In vitro, cell synchronization and subsequent irradiation of cells in sensitive or tolerant phases of the cell cycle are effective. However, cell synchronization is much more complex in animal models, where less promising results were obtained, and so far none of the clinical studies showed any benefit of this approach.

## Classical Radiobiological Research – Future Aspects

There are still open questions which must be addressed by classical radiobiological research. One of these definitely is the question of *consequential late effects*. It was demonstrated that in tissues whose surface is subject to mechanical and/or chemical stress, such as the intestine, urinary bladder, oral mucosa or, to a lesser extent, skin, the acute response of the epithelial lining can markedly influence late effects in these organs. For example, it was demonstrated in the mouse urinary bladder that a marked acute response to radiation, defined as a reduction in individual storage capacity by at least 50%, significantly increased the risk of a late functional response [5, 6, 33]. As the acute bladder response is not associated with urothelial denudation, impairment of the urothelial barrier was postulated as one pathogenetic principle. This was proven by the instillation of heparin or pentosane polysulfate, which both restore the urothelial glycosamino-glycan layer, a major constituent of the barrier [21]. This treatment resulted in a highly significant reduction in acute changes. Moreover, the treatment – carried out during the acute response phase – almost eliminated late functional changes, which normally occur about 6 months after irradiation. This is evidence for a marked consequential component in the urinary bladder response.

Further in vivo experimentation using relevant animal models and focussing on relevant endpoints is required in order to clarify the mechanisms underlying the interaction between the acute and late response in detail. Also, similar effects in radiotherapy patients (e.g. after irradiation for malignancies of the prostate) must be identified.

Another subject, which must be included in translational research in classical radiobiology, is the validity of the assumption of *equal effect per fraction* during the overall treatment time. It has recently been demonstrated in a number of studies that the fractionation effect may change during a course of radiotherapy given over several weeks [7, 24, 25, 32]. A loss of the fractionation sensitivity was reported, indicating that sparing, at least of acutely responding normal tissues by dose fractionation may be lost during the overall treatment time. This issue must be studied qualitatively and quantitatively in relevant animal models to include both time course and extent of the changes. These investigations must aim at identifying the (tissue-specific?) rules followed by the changes. This will then allow to incorporate these results into dose-planning procedures of radiotherapy. Moreover, the efficacy of repopulation processes must be reconsidered, as repopulation must also compensate for the reduced tissue sparing by fractionation.

*Hypersensitivity at low doses per fraction*, and the phenomenon of induced recovery processes, which has been demonstrated for numerous cell lines in vitro [19, 20, 27], but also e.g. for human skin [17, 31], has to be studied for clinically relevant endpoints in animal models. This is of particular relevance for modern radiotherapy protocols with irradiation through multiple fields, where – with a tumor dose per fraction of 2 Gy – a substantial volume of normal tissues is treated with only a small part of this dose per fraction, well below 1 Gy.


## Tissue Radiopathology – Translational Implications

Recently, the classical pathogenetic view of acute radiation reactions, which was strongly influenced by cellular radiobiology, has been extended to a model of *tissue radiopathology*, which includes *dynamic processing* of the reactions after the radiation injury instead of only passive development of hypoplasia and cell depletion [3, 22]. Active processing of radiation damage includes numerous factors, such as changes in signal transduction, intercellular communication (also between different cell populations), protein expression, expression of growth factors. Here, molecular biology approaches have resulted in a major input into the translational chain.

One example, where translational research was effective, are acute radiation effects in *bone marrow*. In this tissue, identification of individual cell subpopulations of the entire hierarchy, from stem cells to multipotential and committed progenitor cells into mature (blood) cells of the individual lineages, is possible via cell surface antigens [18]. Moreover, cells are more readily available than from most other organs. This led to early studies, mostly in vitro, of the factors which influence proliferation and differentiation at the individual stages of cellular development. Today, a number of growth factors, like G-SCF, GM-CSF, erythropoietin, or interleukin-11, are available for treatment of bone marrow toxicity of radiation exposure [15, 18].

Other approaches for the specific intervention aiming at modification of acute radiation effects are based on either blocking of negative signals, or stimulation or substitution of positive signals. These approaches include growth factors, growth factor receptors, modulators of prostaglandin metabolism, like

COX-II inhibitors or essential fatty acid diets, inhibitors of angiotensin-1-converting enzyme, or angiotensin-2 receptor blockers.

*Keratinocyte growth factor* (KGF) is normally produced by fibroblasts, with keratinocytes as effector cells, and hence is an example for a modification of tissue damage by communication between different cell populations. On the basis of in vitro experiments, KGF has been suggested as an agent for selective modulation of acute radiation effects in normal squamous epithelia [23]. A number of in vivo studies have proven the therapeutic potential of this particular growth factor, which is not associated with any toxicity [11, 12, 14, 23]. During daily fractionated irradiation of mouse tongue mucosa, 30–100% – dependent on the administration protocol – of the dose of the first treatment week was found to be compensated by KGF [11].


## A Functional Translational Research Chain

One major prerequisite for a suitable and effective intervention in the development of radiation effects is detailed knowledge of the individual processing steps.

In conclusion, it is essential to combine classical radiobiological results, for each tissue and organ at risk, with modern immunohistochemical and molecular biological studies. On this basis, ways for the intervention in damage processing, e.g. via blocking antibodies to counteract overexpressed signals, or substitution for down-regulated signals, can be developed. Standard in vitro investigations must be supplemented by studies in more complex in vitro models, like coculture of different cell populations or organotypic cultures with relevant substrates. In vivo studies, with suitable animal models of normal tissue responses, and focussing on clinically relevant endpoints, must follow these in vitro investigations. In parallel, the selectivity of possible interventions must be proven by preclinical investigations in established rodent and human tumor models.


## References


1   Baumann M, Bentzen SM, Dörr W, et al. The translational research chain: Is it delivering the goods? Int J Radiat Oncol Biol Phys 2001;49:345–351.
2   Dische S, Saunders MI, Barrett A, et al: A randomised multicentre trial of CHART versus conventional radiotherapy in head and neck cancer. Radiother Oncol 1997;44:123–136.
3   Dörr W: Radiobiological models of normal tissue reactions. Strahlenther Onkol 1998;174 (suppl III):4–7.
4   Dörr W: Three A's of repopulation during fractionated irradiation of squamous epithelia: Asymmetry loss, acceleration of stem-cell divisions and abortive divisions. Int J Radiat Biol 1997;72:635–643.

5   Dörr W, Beck-Bornholdt HP: Radiation-induced impairment of urinary bladder function in mice: Fine structure of the acute response and consequences on late effects. Radiat Res 1999;151:461–467.
6   Dörr W, Bentzen SM: Late functional response of mouse urinary bladder to fractionated X-irradiation. Int J Radiat Biol 1999;75:1307–1315.
7   Dörr W, Brankovic K, Hartmann B: Repopulation in mouse oral mucosa: Changes in the effect of dose fractionation. Int J Radiat Biol 2000;76:383–390.
8   Dörr W, Eckhardt M, Ehme A, Koi S: Pathogenesis of acute radiation effects in the urinary bladder. Experimental results. Strahlenther Onkol 1998;174(suppl III):93–95.
9   Dörr W, Emmendörfer H, Haide E, Kummermehr J: Proliferation equivalent of 'accelerated repopulation' in mouse oral mucosa. Int J Radiat Biol 1994;66:157–167.
10  Dörr W, Kummermehr J: Accelerated repopulation of mouse tongue epithelium during fractionated irradiations or following single doses. Radiother Oncol 1990;17:249–259.
11  Dörr W, Lacmann A, Noack R, Spekl K, Rex K, Farrell CL: Modulation der Strahlenwirkung an Plattenepithelien durch Keratinozyten-Wachstumsfaktor (KGF); in Heinemann G, Müller W-U (eds): Strahlenbiologie und Strahlenschutz. Individuelle Strahlenempfindlichkeit und ihre Bedeutung für den Strahlenschutz. Köln, TÜV-Verlag, 2000, vol 1, pp 209–222.
12  Dörr W, Noack R, Spekl K, Farrell CL: Modification of oral mucositis by keratinocyte growth factor: Single radiation exposure. Int J Radiat Biol 2001;77:341–347.
13  Dörr W, Trott K-R: Strahlenbiologie der Normalgewebe; in Dörr W, Zimmermann JS, Seegenschmiedt MH (eds): Nebenwirkungen in der Radioonkologie. München, Urban & Vogel, 2000, pp 9–23.
14  Farrell CL, Bready JV, Rex KL, et al: Keratinocyte growth factor protects mice from chemotherapy- and radiation-induced gastrointestinal injury and mortality. Cancer Res 1998;58:933–939.
15  Ganser A: Stellenwert hämatopoetischer Wachstumsfaktoren in der Behandlung der hämatopoetischen Insuffizienz nach Strahlenexposition: in Kaffenberger W, Van Beuningen D (eds): Massnahmen in den ersten Tagen nach Strahlenexposition. Aachen, Shaker-Verlag, 1998, pp 72–80.
16  Hall EJ: Radiobiology for the radiobiologist. Philadelphia, Lippincott, 1994.
17  Hamilton CS, Denham JW, O'Brien M, et al: Underprediction of human skin erythema at low doses per fraction by the linear quadratic model. Radiother Oncol 1996;40:23–30.
18  Heyworth CM, Testa NG, Buckle AM, Whetton AD: Growth factors and the regulation of haemopoietic stem cells; in Potten CS (ed): Stem Cells. London, Academic Press, 1997, pp 423–445.
19  Joiner MC: Induced radioresistance: An overview and historical perspective. Int J Radiat Biol 1994;65:79–84.
20  Joiner MC, Lambin P, Malaise EP, et al: Hypersensitivity to very-low single radiation doses: Its relationship to the adaptive response and induced radioresistance. Mutat Res 1996;358:171–183.
21  Krumsdorf D, Noack R, Dörr W: «Consequential late effects» in der Harnblase – experimentelle Ergebnisse. Strahlenther Onkol 1999;175(suppl 1):50.
22  Michalowski A: The pathogenesis and conservative management of radiation injuries. Neoplasma 1995;42:289–292.
23  Ning S, Shui C, Khan WB, Benson W, Lacey DL, Knox SJ: Effects of keratinocyte growth factor on the proliferation and radiation survival of human squamous cell carcinoma cell lines in vitro and in vivo. Int J Radiat Oncol Biol Phys 1998;40:177–187.
24  Rezvani M, Hopewell JW, Morris GM, et al: Repair, repopulation and cell cycle redistribution in rat foot skin. Radiother Oncol 1998;46:193–199.
25  Ruifrok A, Mason K, Hunter N, Thames HD: Changes in radiation sensitivity in mouse skin during fractionated and prolonged treatments. Radiat Res 1994;139:334–343.
26  Scholz M, Kellerer AM, Kraft-Weyrather W, Kraft G: Computation of cell survival in heavy ion beams for therapy – the model and its approximation. Radiat Environ Biophys 1997;36:59–66.
27  Short SC, Mitchell SA, Boulton P, Woodcock M, Joiner MC: The response of human glioma cell lines to low-dose radiation exposure. Int J Radiat Biol 1999;75:1341–1348.
28  Steel GG, McMillan TJ, Peacock JH: The 5Rs of radiobiology. Int J Radiat Biol 1989;56:1045–1048.
29  Thames HD, Hendry JH: Fractionation in Radiotherapy. London, Taylor & Francis, 1987.
30  Thames HD, Bentzen SM, Turesson I, Overgaard M, Van den Bogaert W: Fractionation parameters for human tissues and tumors. Int J Radiat Biol 1989;56:701–710.

31 Turesson I, Flogegard M, Johansson K-A, Svensson P, Wahlgren T, Nyman J: Use of in situ assays and molecular markers to measure individual human tissue response to radiotherapy. Evidence of low dose hypersensitivity and proliferation dependent radiosensitivity. Int J Radiat Oncol Biol Phys 2000;46:737.

32 Turesson I, Thames HD: Repair capacity and kinetics of human skin during fractionated radiotherapy: Erythema, desquamation, and telangiectasia after 3 and 5 year's follow-up. Radiother Oncol 1989;15:169–188.

33 Wiegel T, Dörr W, Hanfmann B: Niere und harnableitende Organe; in Dörr W, Zimmermann JS, Seegenschmiedt MH (Hrsg): Nebenwirkungen in der Radioonkologie. München, Urban & Vogel, 2000, pp 191–198.

34 Withers HR: The four Rs of radiotherapy; in Lett JT, Adler H (eds): Advances in Radiation Biology. New York, Academic Press, 1975, pp 241–271.

Dr. W. Dörr, Klinik und Poliklinik für Strahlentherapie und Radioonkologie,
Medizinische Fakultät Carl Gustav Carus, Technische Universität Dresden,
Fetscherstrasse 74, D–01307 Dresden (Germany)
Tel. +49 351 458 3390, Fax +49 351 458 4347, E-Mail doerr@rcs.urz.tu-dresden.de

Dörr W, Engenhart-Cabillic R, Zimmermann JS (eds): Normal Tissue Reactions in Radiotherapy and Oncology. Front Radiat Ther Oncol. Basel, Karger, 2002, vol 37, pp 9–16

# Experimental Radiotherapy of Late-Responding Tissues – Recent Advances and Future Development

*Klaus-Rüdiger Trott*

St. Bartholomew's and the Royal London School of Medicine and Dentistry, Queen Mary & Westfield College, London, UK

## The Cellular Interpretation of the Pathogenesis of Chronic Radiation Damage

The first clinical study on chronic normal tissue damage was published by Holthusen [8] in 1936. He described the frequency of patients with significant telangiectasia after radiotherapy with different doses as a sigmoid dose-effect relationship and interpreted this curve as reflecting the heterogeneity of the patient's response to primary radiation damage.

By confronting this dose-effect curve of chronic normal tissue damage with the dose-effect curve of local control of skin cancer, he developed the concept of optimal radiation dose for uncomplicated cure. This concept had been forgotten for nearly half a century before it became the bread and butter diet of present-day radiotherapy trainees.

By sheer ingenuity, Holthusen managed to manipulate the two curves to have the same shape and steepness, which is by no means clear from the original data. This similarity may have contributed to a fatal misconception which had a huge influence on radiobiological thinking in the seventies and eighties, namely that the shape of the dose-response curves reflected the underlying mechanisms rather than mere heterogeneity of patient response.

The mechanism of tumour response is well established. It is solely due to the stochastic inactivation of tumour stem cells. Upon completion of radiotherapy, the further fate of the tumour is determined solely by whether or not one or more tumour stem cells have survived [10]. Neither immunological nor other biological processes which might occur after radiotherapy alter this result.

In this sense, tumour cure is a classical deterministic effect. The success of this simple concept to explain the therapeutic results of curative radiotherapy of cancer has persuaded numerous radiobiologists to interpret chronic normal tissue damage by analogy. The inactivation or survival of stem cells or tissue-rescuing units would determine the occurrence or absence of chronic normal tissue damage [21]. Results of fractionation studies were interpreted in terms of shapes of hypothetical stem cell survival curves, and the resulting low $\alpha/\beta$ values were assumed to be the consequence of curvy target cell survival curves. Efforts were made to relate the calculated number of tissue-rescuing units to anatomical structures in the respective organ, such as nephrons or bronchioli [24]. This concept of independent inactivation of functional units and their functional organisation is the basis of the most popular mathematical models of normal tissue complication probability. These are fascinating models which have led to superb experimental and clinical research. Yet, is there any good evidence that this model represents the biology of late responses of irradiated normal tissues? Is it true that these late responses are determined similarly to local tumour control only by the inactivation of some independently responding units of regeneration, be it stem cells or tissue-rescuing units? Differences between patients would then predominantly reflect different numbers of surviving tissue-rescuing units as a consequence of differences in intrinsic cellular radiosensitivity.

## Kinetics of Chronic Radiation Damage Progression

Few radiobiologists did not subscribe to this biophysical concept of chronic normal tissue damage and stressed the importance of pathophysiological processes. The most outspoken and the most influential were Rubin and Casarett [19], who maintained that chronic radiation damage was not a deterministic effect but a dynamic process. Chronic radiation injury is progressive over many years. Turesson [22] demonstrated that the severity of telangiectasia in patients given radiotherapy for breast cancer increased for up to 10 years. This distinguishes chronic radiation damage from acute radiation damage, which displays a well-defined manifestation period followed by a period of healing. The severity of the acute peak reaction depends on the biological radiation dose. Nothing of that can be found in chronic radiation damage, no manifestation period, no healing period. There even is no peak reaction that would be dependent on dose, but it is the progression rate that depends on radiation dose. This dynamic nature of the pathogenesis of typical chronic radiation damage is suggestive of a fundamentally different pathogenetic mechanism. Moreover, it suggests that during this dynamic process a wide range of biological processes could interact with this process, accelerating it or slowing it down. It is exactly this dynamic nature of chronic radiation damage that opens up the theoretical possibilities for therapeutic intervention and post-irradiation prophylaxis.

## The Focal Nature of Development of Chronic Radiation Damage

The characteristic pathological features of most chronic radiation damage are ischaemia, atrophy, fibrosis and necrosis. The most puzzling feature of these changes is their focal appearance which is in clear contrast to the random nature of primary radiation damage. This alone is compelling evidence that it is not the primary radiation damage to the DNA and the clonogenic capacity of 'stem' cells that determines the development of chronic radiation damage to a tissue or an organ. Post-irradiation processes, which involve intercellular communication, determine the final clinical outcome to a large degree.

Myocardial necrosis after irradiation of the heart starts in small foci which enlarge over time [20]. They are not related to the anatomical distribution of blood vessels but closely related to the focal appearance of functional changes in endothelial cells which are manifested within a few days after irradiation, long before any signs of cell death are visible. A similar focal distribution of white matter necrosis has been described in the irradiated central nervous system [23]. Again, this is not related to any pattern of vascular supply. Although there is evidence that white matter necrosis is related to proliferative changes in the oligodendrocyte population, the focal distribution of damage cannot be explained by any direct antiproliferative radiation effect to individual cells, but must be due to multicellular, functional tissue effects, involving intercellular communication.

## Role of Intercellular Communication in the Pathogenesis of Chronic Radiation Damage

In recent years, a great number of studies have focussed on the elucidation of intercellular communication in the pathogenesis of chronic radiation damage. Most studies concentrated on the role of the fibrogenic cytokine TGF-$\beta$. Apparently, it plays a central role in the processing both of acute and of chronic radiation damage. Most published studies investigated the kinetics of TGF-$\beta$ expression in irradiated organs over a period of many months in relation to the development of radiation-induced inflammatory reactions and in particular of radiation fibrosis. In mouse skin, two peaks of TGF-$\beta$ expression were observed [15]. The first was related to the phase of acute radiodermatitis which cleared together with the acute clinical signs. Several months later, TGF-$\beta$ expression increased again in parallel with the beginning histopathological evidence of dermal fibrosis. This suggests that, although TGF-β promotes the fibrotic reaction which finally leads to the characteristic feature of radiation fibrosis, its expression is not directly stimulated by radiation, but is a secondary reaction to other radiation-induced changes.

Similarly, increased TGF-β expression was observed in rats and mice which develop radiation enteropathy [16]. In surgical samples from patients operated on for radiation enteropathy, increased TGF-β immunoreactivity was closely associated with a reduction in thrombomodulin expression in endothelial cells. The authors [17] attributed the rise in TGF-β to the changes in the coagulation mechanisms which appears to be part of the primary chronic radiation effect on the microvascular system. Yet, signalling and modulation goes both ways: increased TGF-β also further down-regulates thrombomodulin expression. Of particular interest is the observation that these changes are focal (just as the functional changes in endothelial cells of the irradiated rat heart are focal [20]). This suggests that focal, functional damage of endothelial cells in the microvasculature is an early step in the complex pathogenesis of chronic radiation damage. I cannot imagine how single-cell inactivation could lead to those focal, functional changes in groups of capillary endothelial cells. Clearly, intercellular communication must play a decisive role.

TGF-β expression is also increased in the small bowel already during the acute, inflammatory phase of radiation enteropathy. A similar early increase of TGF-β has been observed within a few days in a number of other organs, such as mammary gland [2] and lung [6], which do not develop clinically manifest acute inflammatory reactions.

A late increase in TGF-β expression has also been implicated in the development of chronic radiation damage in the bladder [9]. It was demonstrated that the up-regulation of TGF-β expression in the kidney [4] is not a primary radiation effect but a secondary effect modulated by angiotensin.

Of particular interest is the observation that the tendency to develop late radiation fibrosis of the lung [1] and the breast [11] might be identified in the individual patient already by determining TGF-β concentration in a pre-irradiation plasma sample.

These studies suggest that to a large extent radiation fibrosis is driven by the up-regulation of TGF-β. Also the induction of premature differentiation of immature fibroblasts by radiation, which has been observed in vitro [19], may largely be due to the action of increased TGF-β levels on those immature fibroblasts in vivo [3]. This up-regulation is not a direct radiation response, but is secondary to other, progressively deleterious changes in the irradiated tissue probably related to microvascular endothelial cell injury and part of a general protective mechanism characteristic of the involved organ. These observations on the development of chronic radiation injury made during the last few years in various organs also suggest promising potential targets for prophylactic and therapeutic interventions in the future.

## Genetic Susceptibility to Chronic Radiation Damage

All studies described so far (and many more not mentioned here) were designed to identify only the temporal pattern of the expression of a cytokine or another signalling molecule in the course of the development of chronic radiation damage. Yet those studies cannot prove their causative role in the process. Such proof, however, can come from experiments which use strains of experimental animals which differ in the expression of a particular signalling molecule either spontaneously or as a result of genetic modification (knock-out mice).

Clinical experience has taught radiotherapists that patients differ considerably in their genetic susceptibility to develop chronic radiation damage. Radiobiologists have spent great efforts to relate organ sensitivity to genetically determined cellular, intrinsic radiosensitivity. However, there is evidence that even greater interpatient heterogeneity applies to the post-irradiation processing of primary radiation damage. The best studied example for the genetic susceptibility to post-irradiation modification of the development of radiation fibrosis are the studies on mouse strain differences of radiation-induced lung fibrosis. There are quantitative and qualitative differences in radiation pneumotoxicity between mouse strains [7]. C57 mice are prone to radiation-induced hyaline membrane and pulmonary fibrosis. In contrast, CBA and C3H mice exhibit neither of these lesions. The same strain difference was also observed for bleomycin-induced lung fibrosis. This demonstrates that the clinical response is defined more by the genetically determined particular response pattern of the organism than by the particular toxic agent. This difference in susceptibility to lung fibrosis appears to be related to the intrinsic activity of lung plasminogen activator and angiotensin-converting enzyme. The genetic basis of these differences was assessed by crosses and back-crosses between the sensitive and the resistant mouse strains. The genetic analysis suggested that the fibrosis-prone phenotype was controlled by two autosomal dominant genes [7]. Dileto and Travis [5] related the different genetic susceptibility of C57 mice to radiation-induced lung fibrosis to the intrinsic radiosensitivity of lung fibroblasts of both animal strains and concluded that in vitro radiosensitivity of lung fibroblasts as assessed by survival at 2 Gy does not correlate with the development of lung fibrosis in this mouse model.

A similar approach has been successful in further elucidating the pathogenesis of late radiation enteropathy [25]. Mice which are deficient in mast cells showed an exacerbated mucosal injury but minimal reactive fibrosis after irradiation. These findings suggest that mucosal mast cells play a critical role in protecting the intestinal mucosa after initial epithelial injury. If this first-line defence mechanism fails, recruitment of connective tissue mast cells may serve the purpose of reinforcing the intestinal wall by promoting connective tissue deposition. This further supports the conclusion made above that radiation fibrosis is a secondary protective mechanism which is specific to the respective tissue but not to the primary radiation damage.

## Pharmacological Modulation of the Development of Chronic Radiation Damage

Besides those studies on the molecular pathogenesis of chronic radiation damage which use well-defined or genetically modified animals such as knock-out mice, another approach which may shorten the path to the desired clinical application is by modulating damage progression by post-irradiation treatment with specific drugs. Michalowski [12, 13] summarised the evidence for the potential value of this approach, quoting several hundred studies already in the early nineties.

New drugs and new concepts which could serve this purpose are continuously being developed. Most radiobiologists see a particularly promising approach in giving cytokines after irradiation. This treatment is well established for radiation-induced acute bone marrow failure, e.g. using G-CSF or thrombopoietin. However, there is also experimental evidence that the severity and the progression rate of chronic radiation damage can be influenced by cytokines. A particularly fascinating example is the action of basic FGF and PDGF given after irradiation of the spinal cord which has been shown to reduce the incidence and to prolong the latency of radiation myelitis [14].

In addition to stimulation of regeneration, another promising approach is to decrease the chronic inflammation, a characteristic feature of most chronic radiation damage and an important mechanism by which damage progression is aggravated. Different methods have been tested in various experimental models, often with striking success.

## Conclusion

Many excellent models of chronic normal tissue damage for a variety of critical organs and tissues in different experimental animals have been developed over the last decades which permit precise quantification of the severity and incidence of normal tissue damage after irradiation. These models have been used to study the influence of dose fractionation and, to a lesser degree, the effect of volume on the incidence of severe chronic radiation damage, in a purely phenomenological way. The impact of these studies on the daily practice of radiotherapy has been considerable. In addition, some work was devoted to the description of the kinetics of histopathological changes and their relationship to impairment of organ function. I feel that these studies have reached their natural conclusion. In future studies, these experimental models should serve a new and important purpose. A new generation of radiobiological experiments should be designed to study the molecular pathogenesis of chronic radiation dam-age and develop strategies of prevention or treatment of chronic radiation damage after completion of radiotherapy. This research requires, in addition to the techniques of advanced molecular biology, experimental models which permit precise quantification of the clinical response of the living animal. Such studies cannot be performed in vitro. In vitro studies such as those on the effect of irradiation of immature fibroblasts on fibroblast differentiation [18] may suggest potential mechanisms of radiation fibrosis. Yet, interventional studies in vivo are required to test any hypothetical mechanism in the living animal. New therapeutic and preventive strategies can be developed only following a better understanding of the process of chronic normal tissue damage. These interventions could be either genetic, i.e. comparing damage processing and damage progression in different mouse strains or in animals with defined mutations, e.g. in knock-out mice, or by interfering with suspected critical signalling molecules (e.g. cytokines) using specific therapeutic antibodies (e.g. anti-TGF-β), or by specific pharmacological interference with contributing factors, such as chronic inflammation, or by protection against other contributing factors, such as secondary mechanical or chemical injury. Chronic radiation damage is a problem of the organism, not of cells. Only good animal experiments will be able to lead to substantial progress in the clinical management of this harrowing problem of cancer therapy. This has to be based on good experimental science rather than on empirical observation.

## References


1   Anscher MS, Kong FM, Jirtle RL: The relevance of transforming growth factor beta 1 in pulmonary injury after radiation therapy. Lung Cancer 1998;19:109–120.
2   Barcellos-Hoff MH, Derynck R, Tsang MI, Weatherbee JA: Transforming growth factor-β activation in irradiated murine mammary gland. J Clin Invest 1994;93:892–899.
3   Burger A, Löffler H, Bamberg M, Rodemann HP: Molecular and cellular basis of radiation fibrosis. Int J Radiat Biol 1998;73:401–408.
4   Datta PK, Moulder JE, Fish BL, Cohen EP, Llanos EA: TGF-β1 production in radiation nephropathy: Role of angiotensin II. Int J Radiat Biol 1999;75:473–479.

5    Dileto CL, Travis EL: Fibroblast radiosensitivity in vitro and lung fibrosis in vivo: Comparison between a fibrosis-prone and fibrosis-resistant mouse strain. Radiat Res 1996;146:61–67.

6    Finkelstein JN, Johnston CJ, Baggs R, Rubin P: Early alterations in extracellular matrix and transforming growth factor-β gene expression in mouse lung indicative of late radiation fibrosis. Int J Radiat Oncol Biol Phys 1994;28:621–631.

7    Franco AJ, Sharplin J, Ward WF, Taylor JM: Evidence for two patterns of inheritance of sensitivity to induction of lung fibrosis in mice by radiation, one of which involves two genes. Radiat Res 1996;146:68–74.

8    Holthusen H: Erfahrungen über die Verträglichkeitsgrenze für Röntgenstrahlen und deren Nutzanwendung zur Verhütung von Schäden. Strahlentherapie 1936;57:254–269.

9    Kraft M, Oussoren Y, Stewart FA, Dörr W, Schultz-Hector S: Radiation-induced changes in transforming growth factor-β and collagen expression in the murine bladder wall and its correlation with bladder function. Radiat Res 1996;146:619–627.

10   Kummermehr J, Trott KR: Tumour stem cells; in Potten CS (ed): Stem Cells. London, Academic Press, 1997, pp 363–399.

11   Li C, Wilson PB, Levine E, Barber J, Stewart AL, Kumar S: TGF-β1 levels in pre-treatment plasma identify breast cancer patients at risk of developing post-radiotherapy fibrosis. Int J Cancer 1999;84:155–159.

12   Michalowski A: On radiation damage to normal tissues and its treatment. Growth factors. Acta Oncol 1990;29:1017–1023.

13   Michalowski A: Anti-inflammatory drugs. Acta Oncol. 1994;33:139–157.

14   Nieder C, Ataman F, Price RE, Ang KK: Biologische Präventions- und/oder Therapiemöglichkeiten der Strahlenmyelopathie. Strahlenther Oncol 1999;175:437–443.

15   Randall K, Coggle JE: Long-term expression of transforming growth factor TGF-β1 in mouse skin after localized β-irradiation. Int J Radiat Biol 1996;70:351–360.

16   Richter KK, Langberg CW, Sung CC, Hauer-Jensen M: Association of transforming growth factor (TGF-β) immunoreactivity with specific histopathologic lesions in subacute and chronic experimental radiation enteropathy. Radiother Oncol 1996;39:243–251.

17   Richter KK, Fink LM, Highes BM, Sung CC, Hauer-Jensen M: Is the loss of endothelial thrombomodulin involved in the mechanism of chronicity in late radiation enteropathy. Radiother Oncol 1997;44:65–71.

18   Rodemann HP, Bamberg M: Cellular basis of radiation fibrosis. Radiother Oncol 1995;35:37–43.

19   Rubin P, Casarett GW: Clinical Radiation Pathology. Philadelphia, Saunders, 1968, vol 1, 2.

20   Schultz-Hector S: Heart; in Scherer E, Streffer C, Trott KR (eds): Radiopathology of Organs and Tissues. Berlin, Springer 1991, pp 347–368.

21   Thames HD, Hendry JH: Fractionation in Radiotherapy. London, Taylor & Francis, 1987

22   Turesson I: The progression rate of late radiation effects in normal tissues and its impact on dose–response relationships. Radiother Oncol 1989;15:217–226.

23   Van der Kogel A: The nervous system: Radiobiology and experimental pathology; in Scherer E, Streffer C, Trott KR (eds): Radiopathology of Organs and Tissues. Berlin, Springer, 1991, pp 191–212.

24   Withers HR, Taylor JM, Maciejewski B: Treatment volume and tissue tolerance. Int J Radiat Oncol Biol Phys 1988;14:751–759.

25   Zheng H, Wang J, Sung CC, Hauer-Jensen M: Protective role of mast cells in intestinal radiation toxicity (abstract NTR 16). 11th Int Congr Radiat Res, Dublin 1999.



Prof. K.R. Trott, Gray Laboratory, Mount Vernon Hospital,
HA6 2JR Northwood (UK)
Tel. +44 192 382 8611, Fax +44 498 022 7287, E-Mail k.r.trotte@mds.qmw.ac.uk

Dörr W, Engenhart-Cabillic R, Zimmermann JS (eds): Normal Tissue Reactions in Radiotherapy
and Oncology. Front Radiat Ther Oncol. Basel, Karger, 2002, vol 37, pp 17–25

# Functional Evaluation of the Human Pancreas before and in the Early Period after Hyperfractionated Accelerated Radiochemotherapy

*Eckehard Horst*[a], *Matthias Seidel*[b], *Oliver Micke*[a], *Christian Rübe*[a],
*Ulrich Schäfer*[a], *Normann Willich*[a]

[a] Department of Radiation Oncology and [b] Department of Medicine,
University of Münster, Germany

Pancreatic cancer is the second commonest gastrointestinal cancer in Europe and most other North American countries. The incidence is estimated to be around 8–9 per 100,000. Due to an unspecific symptomatology, the majority of patients present with locally advanced disease with involvement of regional lymph nodes, adjacent vessels and other tissues. 5-fluorouracil (5-FU)-based radiochemotherapy has been established as an effective treatment option in this patient population. It is effective in achieving durable local control and alleviating pain, and median survival was demonstrated to be 9–12 months in comparison to 3–5 months in patients receiving palliative care only [18, 21]. During the course of their disease, most patients complain of symptoms like anorexia, weight loss, diarrhea and meteorism. This symptomatology of gastrointestinal dysfunction can be caused by tumor progression, but may also be influenced by therapeutic procedures. The result is a condition of catabolism and malnutrition which is generally recognized as a factor with negative influence on survival and quality of life [12]. Also, resection of pancreatic tissue obviously can result in functional impairment [16, 19], and exocrine and endocrine pancreatic dysfunctions can contribute to nutritional depletion.

During radiotherapy, parts of the pancreatic gland are exposed to high radiation doses. At present, evidence is available mainly on histopathologic changes in the human pancreas after exposure to ionizing radiation [23], while data on exocrine function are scarce, because direct test procedures are invasive and technically difficult [14, 17]. Studies on serum enzyme level have demonstrated low specificity. However, recently developed indirect assays based on the protein-synthesizing capacity of pancreatic glandular tissue and on the identification of pancreatic elastase 1 in feces allow the quantitation of exocrine pancreatic function [2, 6, 8, 9]. In the present study, we evaluated biochemical endpoints for exocrine and endocrine pancreatic functions in patients with stage III/IVa pancreatic carcinoma immediately prior to and early after hyperfractionated accelerated radiochemotherapy (HFRT). The results are compared with published experimental data.

## Patients and Methods

### Patients

Sixteen patients undergoing radiochemotherapy for locally advanced pancreatic adenocarcinoma at Münster University Hospital were selected. Patients with a history of chronic pancreatitis, diabetes mellitus or liver dysfunction were not included. Five patients were excluded in the course of the study because of one or more of the following reasons: lack of compliance, acute pancreatitis, jaundice, clinical deterioration and/or tumor progression. The latter was indicated by an increase in CA19-9 and/or CEA values, or detected with CT, endoscopic retrograde cholangiopancreatography and endosonography. Patient characteristics of the 11 patients available for evaluation are shown in table 1.

### Hyperfractionated Accelerated Radiochemotherapy

Three-dimensional conformal therapy with photons of 10 or 15 MV was applied. A total dose of 44.8 Gy to the 90% isodose was administered with 1.6 Gy per fraction, given twice daily. The target volumes were the tumor area as defined by CT, MRI and clips after exploration and the nodal areas at risk. The entire duodenal loop was included if the tumor was located in the pancreatic head. On the first 3 days of radiotherapy, 600 mg/m$^2$ of 5-FU was administered intravenously as a 10-min short infusion.

Differential dose-volume histograms (DVH) of the entire pancreatic gland were calculated for each patient. For the characterization of the DVHs, mean organ dose ($D_{mean}$), and maximum dose to 10% of organ volume ($D_{10\%}$) were extracted from the DVH data [5]. $D_{10\%}$ was evaluated since at least 10% of the glandular volume are considered necessary to maintain functional capacity [13].

### Function Analysis

*Amino Acid Consumption Test.* The decline in the plasma amino acid level (PAL) upon endocrine stimulation indicates the protein-synthesizing capacity of the pancreatic exocrine glandular tissue. The routine amino acid consumption test (AACT) procedure was performed. Blood samples for PAL determination were taken from fasted patients before (baseline) and at 15-min intervals after a 1-hour intravenous infusion of secretin (1 CU/kg body weight), ceruletide (5 μg) and 1 ml human albumin (20%). Total plasma amino acids, besides proline and hydroxyproline, were determined. The cutoff limit of ≤12% was used as an indicator of functional impairment [6]. The test was performed immediately prior to, and 28 or 56 days after HFRT.

*Elastase Assay.* Elastase concentration was measured with an enzyme-linked immunoassay kit using monoclonal antibodies against different specific epitopes of human pancreatic elastase 1. The cutoff level indicating a decreased exocrine function was 200 μg/g. Elastase

***Table 1.**  Patient characteristics*

| Patient no. | Age/sex | Symptom interval[a] | UICC/ grading | HFRT duration days |
|---|---|---|---|---|
| 1 | 55/f | 17 | IVA/GII | 18 |
| 2 | 57/m | 8 | III/GIII | 19 |
| 3 | 59/f | 10 | III/GII | 20 |
| 4 | 45/m | 4 | IVA/GII | 18 |
| 5 | 72/m | 7 | IVA/GII | 18 |
| 6 | 44/f | 6 | IVA/GII | 21 |
| 7 | 57/m | 13 | III/GII | 18 |
| 8 | 68/f | 9 | III/GIII | 18 |
| 9 | 78/f | 24 | IVA/GII | 21 |
| 10 | 39/f | 5 | IVA/GII | 20 |
| 11 | 73/m | 11 | III/GII | 18 |
| Median | 59 | 10 | | 19 |

[a] Time till diagnosis (weeks).

determination was performed immediately prior to, and 4 or 8 weeks after HFRT. Each time, measurements were performed twice within 5 days of fecal collection. Elastase concentration could not be obtained in 4 patients due to diarrhea.

*Endocrine Function.* Plasma glucose profiles were defined immediately before, at weekly intervals during and 8 weeks after HFRT. Glycosylated hemoglobin A1c (HbA1c) values were measured before and 3 months after HFRT. The release of insulin-connecting peptide (C-peptide) in the fasting state was determined as an index of insulin secretion before and at 1 and 2 months after HFRT [4].

*Clinical Parameters.* Weight loss before, during and after HFRT, and clinical diarrhea/steatorrhea were recorded.

*Statistical Analysis.* The pre-HFRT and post-HFRT results of pancreatic function analysis were compared by Wilcoxon's test for paired differences. Correlation coefficients were calculated between $D_{mean}$, $D_{10\%}$ and exocrine function parameters using the Spearman rank correlation statistic. $p < 0.05$ was considered statistically significant. Data are expressed as median and range.

## Results

### Exocrine Function

The median PAL decline was 17% (10–29) before HFRT and dropped to 10% (5–20) 28–56 days after HFRT (table 2). The relative change of the median PAL decline was 41.2% (p = 0.02). Low values <12% were ascertainable in 3 of 11 patients before HFRT. Seven additional patients developed pathological results in the AACT 28 or 56 days after HFRT, whereas a total of 4 patients had

*Table 2.* DHV statistics, clinical findings and pancreatic function analysis pre- and post-accelerated radiochemotherapy

| Patient no. | $D_{mean}$ Gy | $D_{10\%}$ Gy | Weight[a] kg | Loss % kg | Diarrhea WHO | AACT % | | Elastase 1[b] $10^{-6}$ g/g | |
|---|---|---|---|---|---|---|---|---|---|
| | | | | | | pre | post | pre | post |
| 1 | 44.0 | 20.4 | 62/56 | 9.7 | II | 10 | $14^{II}$ | – | – |
| 2 | 39.3 | 25 | 69/67 | 2.9 | III | 29 | $19^{I}$ | – | – |
| 3 | 39.3 | 16.2 | 65/60 | 7.7 | II | 18 | $20^{II}$ | – | – |
| 4 | 46.4 | 15.7 | 45/43 | 4.5 | II | 17 | $8^{I}$ | – | – |
| 5 | 48.7 | 44.8 | 73/65 | 11.0 | – | 17 | $5^{I}$ | 290 | 180 |
| 6 | 46.9 | 43.6 | 63/63 | 0.0 | – | 13 | $10^{I}$ | 195 | 15 |
| 7 | 39.5 | 21.4 | 78/72 | 7.7 | – | 21 | $8^{I}$ | 240 | 15 |
| 8 | 46.6 | 36.5 | 50/45 | 10.0 | – | 10 | $10^{I}$ | 120 | 100 |
| 9 | 41.8 | 16.2 | 88/88 | 0.0 | – | 25 | $10^{II}$ | 600 | 400 |
| 10 | 43.9 | 17.9 | 65/58 | 11.0 | – | 10 | $5^{II}$ | 18 | 15 |
| 11 | 42.3 | 14.2 | 81/83 | 0.0 | – | 18 | $13^{II}$ | 100 | 85 |
| Median | 43.9 | 20.4 | | 7.7 | | 17 | 10 | 195 | 85 |

[a] Weight before and 28–56 days after HFRT.
[b] Elastase 1 could not be obtained in 4 patients due to diarrhea.

a normal PAL decline after HFRT indicating a preserved glandular capacity. Median intraluminal elastase concentrations declined from 195 µg/g (18–600) to 85 µg/g (15–400, relative change 56.2%, p = 0.02). Clear pathologic values were obtained in 3 patients before HFRT and in an additional 4 patients after HFRT (table 2). Two patients showed normal elastase concentrations after HFRT, indicating a preserved functional capacity. Using the cutoff values for functional impairment of both tests, a discrepancy is noted in patient 9 and 11. In patient 9, it can be attributed to the individual levels of sensitivity and specificity of the methods. The discrepancy in patient 11 can be explained by different pathophysiological reasons for a glandular and functional impairment. For example, functional impairment may be due to a duct occlusion or a disturbed intestinal metabolism [11]. The combined average results indicate a moderate glandular and functional impairment after HFRT. One patient had pathologic PAL decline and elastase values before and improved values after HFRT.

*Endocrine Function*

Normal HbA1c values were required for participation in the study (median 4.6%, range 3.6–5.8%). Median HbA1c was 5.0% 3 months after HFRT, which is slightly higher (p = 0.02), but remained in the normal range (2.7–6.6%).

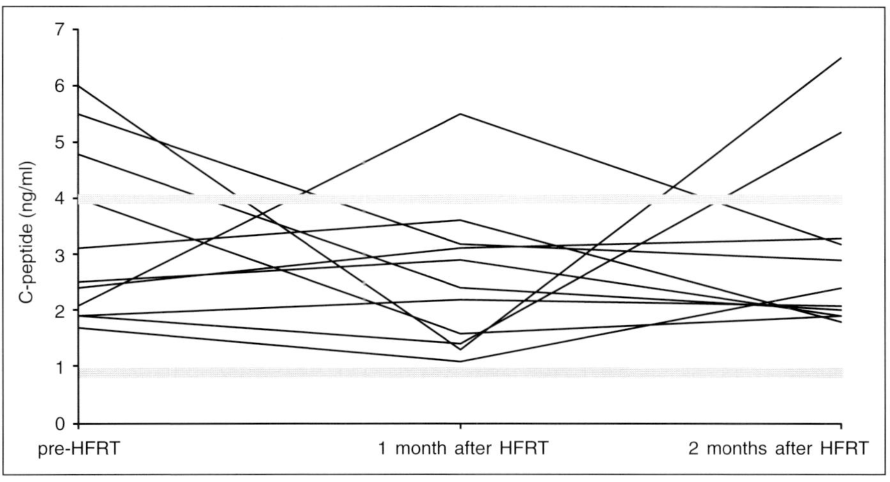



***Fig. 1.*** The gray lines indicate the normal concentration range of C-peptide (0.9–4.0 ng/ml).

Plasma glucose profiles remained in the normal range. No concentrations above 140 mg/dl occurred in the fasting state. Islet β-cell function was estimated by C-peptide levels in the fasting state. The individual results immediately before HFRT and 1 and 2 months after HFRT are shown in figure 1. The majority of values were in the normal range (0.9–4.0 ng/ml). No decreased secretion was indicated during the study. However, a subclinical islet β-cell dysfunction cannot be excluded, because no C-peptide kinetics were obtained [4].

### Clinical Parameters

The median relative weight loss during the observation period was 7.7% (0–11). Diarrhea WHO II occurred in 3 patients and diarrhea WHO III in 1 patient. Steatorrhea was not observed (table 2).

### Differential Dose-Volume Histogram Statistics

Median $D_{mean}$ to the entire pancreatic gland was 43.9 Gy (39.3–48.7). Median $D_{10\%}$ was 20.4 Gy (14.2–44.8), as illustrated in table 2. No significant correlation was found between $D_{mean}$, $D_{10\%}$ and PAL decline or elastase concentration decline.

## Discussion

### Exocrine Component

This study evaluated the influence of HFRT on the secretory exocrine function of the human pancreas 4–8 weeks after therapy. Median $D_{mean}$ to the entire pancreatic gland extracted from the DVH data was 43.9 Gy and median $D_{10\%}$ was 20.4 Gy (14.2–44.8). The radiation effect was augmented by 5-FU. The average findings indicate a moderate glandular and functional impairment associated with HFRT, although the validity of the data and the statistical results have to be confirmed by studies on a larger number of patients. Normal parameters were measured in 3 patients after HFRT. The significance of weight loss and steatorrhea as important clinical symptoms of exocrine pancreatic insufficiency was decreased by the acute gastrointestinal toxicity that occurs during 5-FU-based chemoradiation and the rather short observation period. No correlations between the DVH parameters and exocrine function parameters were observed, which can be explained by the similarity of the total doses applied and the small number of patients. However, conformal treatment techniques may contribute to the preserved glandular capacity measured by the AACT in 4 patients because parts of the gland can be protected, e.g. the pancreatic tail in case of tumors of the pancreatic head. The glandular volume that was shielded from the planned target volume varied between 10 and 40% in this study (data not shown). It must be stressed that the results are affected by the underlying pancreatic carcinoma, which can induce secretory abnormalities by anatomic obstruction and the induction of inflammatory alterations [22, 25]. Two patients had pathologic results in both tests before HFRT. However, clinical findings, blood chemical values including the tumor markers CA19-9 and CEA, and diagnostic imaging studies performed on the study population showed no evidence of tumor progression or acute pancreatitis during the observation period.

*Experimental Data*

The effect of ionizing radiation on the integrity of the pancreatic structure and function has been investigated in canine models. Several experiments describe the gross morphologic and histopathologic alterations [3, 24, 26, 27]. The results of experiments using megavoltage equipment and a dose range comparable to the dose applied in the treatment of pancreatic cancer show an induction of fibrosis of pancreatic tissue and a progressive loss of normal acinar cells 7–8 weeks after irradiation [1, 20]. The few experimental studies which investigated exocrine function have yielded inconclusive results. Different fraction sizes and time intervals and different irradiated volumes that were used in the laboratory have to be considered. Pieroni et al. [20] studied short- and long-term effects of 14 days of $^{60}$Co-radiotherapy at a dose of $6 \times 4$ Gy on the exocrine pancreas in 6 dogs. Pancreatic secretion was collected via direct cannulation of the pancreatic duct after stimulation with secretin and cholecystokinin-pancreozymin. Following a brief period of slight hypersecretion during the first week of irradiation, a progressive reduction of flow, bicarbonate and enzyme output occurred within the early period 2–5 weeks after irradiation. The output was reduced to a rate of 20% for at least 4 months with no signs of recovery. Heijmans et al. [10] assessed exocrine function after intraoperative radiation therapy with 25 Gy, 30 Gy or 35 Gy by fecal fat excretion, which remained unchanged. However, fecal fat excretion would only detect severe states of exocrine insufficiency with more than 90% reduction in glandular capacity [13].

*Endocrine Component*

In contrast to the functional changes in the exocrine component, no significant disturbance of the endocrine component was found in the short term. No patient developed signs of a diabetic metabolism. Experimental data of studies using canine models support this finding. No evidence for disturbed endocrine function was found with fasting blood glucose and glucose tolerance tests 8 weeks after $^{60}$Co irradiation with 6 × 4 Gy or 135 days after intraoperative radiation therapy at 17.5–40 Gy plus 50 Gy of external beam irradiation [1, 20]. Heijmans et al. [10] investigated the functional tolerance of the endocrine canine pancreas using glucose clearance rates and serum insulin levels. A significant decline in the measured parameters was noted at the 1-year of follow-up after intraoperative radiation therapy at 35 Gy, but there were no signs of overt diabetes. Consistently, histopathologic light-microscopic studies reported a dose-related atrophy of acinar cells and radiation damage to blood vessels, but no significant effects on the islet cells of Langerhans [1]. An electron-microscopic study described reversible ultrastructural alterations up to 1 year after 250-kV X-irradiation (5.000–9,000 R) without any negative functional correlation [26, 27].

## Conclusion

Our data indicate that HFRT in pancreatic adenocarcinoma can be associated with glandular and functional impairment of the exocrine component in the short term. As a therapeutic consequence, lipase-rich enzyme supplementation should be administered, particularly if the patients lose weight despite a sufficient caloric intake. In contrast, the endocrine component appears to be more radioresistant, since no deterioration occurred. Progressive fibrosis may adversely affect both components in the longer run but both complications can be treated effectively.

## References


1   Ahmadu-Suka F, Gillette EL, Withrow SJ, et al: Pathologic response of the pancreas and duodenum to experimental intraoperative irradiation. Int J Radiat Oncol Biol Phys 1988;14:1197–1204.

2    Allen AM, Oates PS: Measuring total plasma amino acid concentrations as a test of exocrine pancreatic function. Gut 1992;33:392–396.

3    Archambeau J, Griem M, Harper P: The effect of 250-kV x-rays on the dog's pancreas. Morphological and functional changes. Radiat Res 1966;28:243.

4    Christiansen E, Tibell A, Groth CG, et al: Limitations in the use of insulin or C-peptide alone in the assessment of β-cell function in pancreas transplant recipients. Transplant Proc 1994;26:467–468.

5    Dale E, Olsen DR: Specification of the dose to organs at risk in external beam radiotherapy. Acta Oncol 1997;36:129–135.

6    Domschke S, Heptner G, Kolb S, et al: Decrease in plasma amino acid level after secretin and pancreozymin as an indicator of exocrine pancreatic function. Gastroenterology 1986;90:1031–1038.

7    Gullo L, Pezzilli R, Ventrucci M, et al: Caerulein-induced plasma amino acid decrease: A simple, sensitive, and specific test of pancreatic function. Gut 1990;31:926–929.

8    Gullo L, Ventrucci, Tomassetti P, et al: Fecal elastase 1 determination in chronic pancreatitis. Dig Dis Sci 1999;44:210–213.

9    Hamwi A, Veitl M, Maenner G, et al: Pancreas elastase 1 in stool: Variation within one stool passage and individual changes from day to day. Wien Klin Wochenschr 2000;112:32–35.

10    Heijmans HJ, Mehta DM, Kleibeuker JH, et al: Intraoperative irradiation of the canine pancreas: Short term effects. Radiother Oncol 1993;29:347–351.

11    Heptner G, Domschke S, Domschke W: Exocrine pancreatic function after gastrectomy. Specificity of indirect tests. Gastroenterology 1989;97:147–153.

12    Klein S, Kinney JM, Jeejebhoy KN, et al: Nutrition support in clinical practice: Review of published data and recommendations for future research directions. Am J Clin Nutr 1997;66:683–706.

13    Lankisch PG, Lembcke B, Wemken G, et al: Functional reserve capacity of the exocrine pancreas. Digestion 1986;35:175–181.

14    Levy P, Menzelxhiu A, Paillot B, et al: Abdominal radiotherapy is a cause for chronic pancreatitis. Gastroenterology 1993;105:905–909.

15    Marghini A, Nelson DK, Jones JD, et al: Is the plasma amino acid consumption test an accurate test of exocrine pancreatic insufficiency? Gastroenterology 1994;106:488–493.

16    McLeod RS, Taylor BR, O'Connor BI, et al: Quality of life, nutritional status and gastrointestinal hormone profile following the Whipple procedure. Am J Surg 1995;169:179–185.

17    Mitchel CJ, Simpson FG, Davison AM, et al: Radiation pancreatitis: A clinical entitiy? Digestion 1979;19:134–136.

18    Moertel CG, Frytak S, Hahn RG, et al and The Gastrointestinal Tumor Study group: Therapy of locally unresectable pancreatic carcinoma: A randomized comparison of high dose (6,000 rads) radiation alone, moderate dose radiation (4,000 rads + 5-fluorouracil), and high dose radiation + 5-fluorouracil: Cancer 1981;48:1705–1710.

19    Ohshio G, Tanaka T, Imamura T, et al: Exocrine pancreatic function in the early period after pancreatoduodenectomy and effects of preoperative pancreatic duct obstruction. Dig Dis Sci 1996;41:1947–1952.

20    Pieroni PL, Rudick J, Adler M, et al: Effect of irradiation on the canine exocrine pancreas. Ann Surg 1976;184:610–614.

21    Prott FJ, Schonekaes K, Preusser P, et al: Combined modality treatment with accelerated radiotherapy and chemotherapy in patients with locally advanced inoperable carcinoma of the pancreas: Results of a feasibility study. Br J Cancer 1997;75:597–601.

22    Rinderknecht H, Renner IG, Stace NH: Abnormalities in pancreatic secretory profiles of patients with cancer of the pancreas. Dig Dis Sci 1983;28:103–110.

23    Schoonbroodt D, Zipf A, Herrmann G, et al: Histological findings in chronic pancreatitis after abdominal radiotherapy. Pancreas 1996;12:313–315.

24    Schultz-Hector S, Brechenmacher P, Dörr W, et al: Complications of combined intraoperative radiation (IORT) and external radiation (ERT) of the upper abdomen: An experimental model. Radiother Oncol 1996;38:205–214.

25    Ventrucci M: Enzymology; in Pederzoli P, Cavallini G, Bassi C (eds): Facing the Pancreatic Dilemma: Update of Medical and Surgical Pancreatology. New York, Springer, 1994.

26    Volk BW, Wellmann KF, Lewitan A: The effect of radiation on the fine structure and enzymes of the dog pancreas. I. Short term studies. Am J Pathol 1966;48:721–752.

27    Wellmann KF, Volk BW, Lewitan A: The effect of radiation on the fine structure and enzyme content of the dog pancreas. II. Long-term studies. Lab Invest 1966;15:1007–1023.


Eckehard Horst, MD, Department of Radiation Oncology, University of Münster,
Albert-Schweitzer-Straße 33, D–48129 Münster (Germany)
Tel. +49 251 83 47831, Fax +49 251 83 47355, E-Mail horste@uni-muenster.de

Dörr W, Engenhart-Cabillic R, Zimmermann JS (eds): Normal Tissue Reactions in Radiotherapy
and Oncology. Front Radiat Ther Oncol. Basel, Karger, 2002, vol 37, pp 26–33

# Comparison of Early Pulmonary Changes in [18]FDG-PET and CT after Combined Radiochemotherapy for Advanced Non-Small-Cell Lung Cancer: A Study in 15 Patients

*Ursula Nestle*[a], *Dirk Hellwig*[a], *Jochen Fleckenstein*[b], *Karin Walter*[d],
*Dieter Ukena*[c], *Christian Rübe*[b], *Carl-Martin Kirsch*[a],
*Michael Baumann*[e]

Departments of [a]Nuclear Medicine, [b]Radiation Therapy and Radiooncology,
Clinic of Radiology and [c]Department of Medicine V, Clinic of Medicine,
University Hospital of the Saarland, Homburg/Saar, [d]Department of Radiation
Therapy, Marienkrankenhaus Amberg, [e]Clinic of Radiation Therapy and Radiation
Oncology, Medical Faculty Carl Gustav Carus, University of Dresden, Germany

The lung is a radiosensitive organ, and therefore an important dose-limiting normal tissue for radiotherapy of intrathoracic cancer. Radiation-induced pneumopathy may occur as an inflammatory reaction (radiation pneumonitis) during the first 3 or 4 months after irradiation, and after several months or years as fibrosis [11]. Although considerable research efforts have been directed to the biological mechanism of radiation-induced pneumopathy [3], prevention is presently the only available method to limit radiation-induced lung morbidity [1].

Radiation-induced pneumopathy usually is diagnosed and monitored by chest X-ray and CT. With these techniques, density changes in the lung that are caused by tissue edema and/or fibrosis are detected [12]. Less frequently, functional imaging methods like lung perfusion and ventilation scintigraphy or gallium scans are used [10, 17]. In recent years, [18]FDG-PET has been increasingly applied in the staging and follow-up of lung cancer [4, 6, 9, 14, 16, 18–20]. [18]FDG-PET monitors cellular glucose metabolism in vivo, which is increased in many tumors. However, inflammatory reactions may lead to [18]FDG enhancement as well, and radiation-induced pneumonitis has been reported to interfere with tumor-induced changes in restaging PET [2, 5]. This observation suggests that [18]FDG-PET might bear a potential for noninvasive research into the mechanisms and the time course of radiation-induced lung damage. The aim of the present retrospective study was to compare the incidence and severity of radiation pneumonitis in [18]FDG-PET and CT in a cohort of 15 patients who received radiation therapy for non-small-cell lung cancer (NSCLC).

## Material and Methods

The data of 15 patients with advanced NSCLC (UICC stage III) (n = 11) and limited stage IV (n = 4) were evaluated. The patients underwent palliative combined sequential radiochemotherapy (primary RCT: n = 14; postoperative RCT: n = 1) with 4 courses of carboplatin (AUC 6) and vinorelbine (30 mg/m$^2$). Accelerated irradiation of the tumor region and the mediastinum was given to a total dose of 32 Gy with two daily fractions of 2 Gy (minimum time interval between fractions: 6 h). Irradiations were given with ap/pa field- (n = 12), 4-field- (n = 2) or oblique-field techniques (n = 1) after CT-based treatment planning. Dose prescription and reporting were performed according to the recommendations of Report 50 of the International Commission on Radiation Units and Measurements (ICRU-50) [8], correction for tissue inhomogeneities was performed.

All patients had a [18]FDG-PET examination before and 2.9 months (median; range: 1.4–3.9 months) after initiation of radiotherapy. The PET emission and transmission images were obtained on an ECAT ART PET scanner with an axial field of view of 16.2 cm (Siemens/CTI) in a multi-bed whole-body technique. Singles transmission scanning was performed with a collimated [137]Cs source and used to correct the emission data for attenuation. Emission scans were acquired 90–150 min after intravenous injection of 250 MBq [18]FDG. Blood glucose levels were measured before injection of [18]FDG, and were below 130 mg/dl in all cases. Images were reconstructed by iterative reconstruction using OSEM [7] in a 128 × 128 matrix and processed to a whole-body volume file.

All patients had spiral contrast-enhanced chest CT scans before therapy and within a median of 3 days (range: 1–7 days) before or after the posttherapeutic [18]FDG-PET examination.

To evaluate the effects of irradiation in PET and CT, four regions of interest (ROIs) were defined in the treatment planning CT (fig. 1).

ROI 1 (tumor): small volume within the tumor mass;

ROI 2 (ipsilateral irradiated lung): ipsilateral lung without pathomorphologic changes within the 80% isodose. In order to avoid spill-over effects, lung tissue directly adjacent to malignant tissue was excluded;

ROI 3 (contralateral irradiated lung): contralateral lung without pathomorphologic changes within the 80% isodose;

ROI 4 (reference lung): lung volume without pathomorphologic changes outside the irradiated volume.

The ROIs defined in the planning CT were transferred to the PET images using anatomical landmarks. For all ROIs, in the pre- and posttherapeutic [18]FDG-PET studies, standardized uptake values (SUVs), i.e. the relative uptake of FDG/pixel compared to an idealized

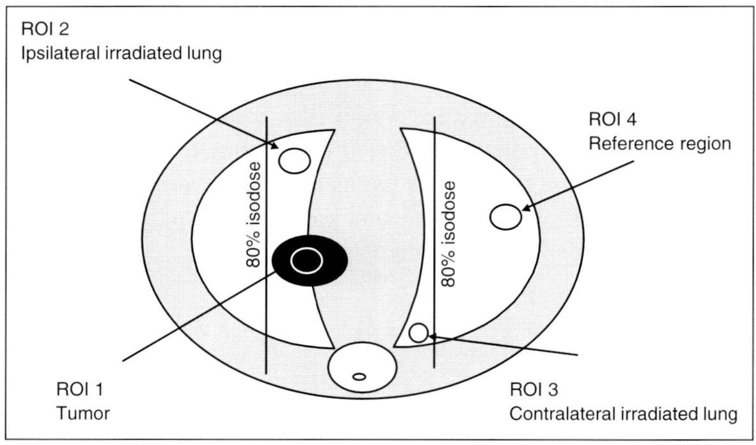


***Fig. 1.*** ROIs defined for analysis in the planning CT of the chest.

body mean [15], were determined. The average ($SUV_a$) and peak SUVs ($SUV_p$) for each ROI were recorded.

By comparison to the pretherapeutic images, the restaging CT scans were reviewed for signs of pulmonary radiation effects within ROIs 2–4, using a 4-step score from 0 (no change) to 3 (dense changes) according to the 'objective' category (grade 0–3 of 'pulmonary fibrosis') of the LENT-SOMA scoring system [21].

Statistical evaluation was done by standard procedures like $\chi^2$ test, Fisher's exact test and analysis of correlation.

## Results

### ROI 1 (Tumor)

Before therapy the median $SUV_a$ of the primary tumors was 8.2 (range: 3.4–20.2), and the median $SUV_p$ was 14.2 (5.5–34.8). After therapy, the median $SUV_p$ of the primary tumors dropped to 4.9 (range: 1.7–16.3) and the $SUV_a$ to 2.0 (range: 1.1–3.7). As expected, the pretherapeutic and posttherapeutic SUVs values varied considerably among the patients (table 1).

### ROIs 2–4 (Lung Tissue)

Before therapy, the SUVs in the three ROIs of lung tissue were significantly lower than the tumor values (median $SUV_a$ = 0.4–0.6; median $SUV_p$ = 0.9–1.3), and did not show significant differences between the lung regions. Unlike the tumor-ROIs, the SUVs in the unirradiated lungs did not show substantial interindividual variation.

***Table 1.*** ROI parameters

| ROI | Patients n | Median volume cm$^3$ | Before therapy | | After therapy | |
|---|---|---|---|---|---|---|
| | | | med. SUV$_a$ | med. SUV$_p$ | med. SUV$_a$ | med. SUV$_p$ |
| ROI 1: tumor | 14[a] | 27 | 8.2 (3.4–20.2) | 14.2 (5.5–34.8) | 2.0 (1.1–3.7) | 4.9 (1.7–16.3) |
| ROI 2: ipsilateral lung | 15 | 57 | 0.6 (0.4–0.8) | 1.3 (0.7–1.8) | 0.9 (0.5–3.0) | 1.8 (1.1–6.5) |
| ROI 3: contralateral lung | 13[b] | 50 | 0.5 (0.3–0.7) | 0.9 (0.7–1.2) | 0.7 (0.4–0.9) | 1.3 (0.9–1.8) |
| ROI 4: reference lung | 15 | 73 | 0.4 (0.3–0.6) | 1.1 (0.6–1.6) | 0.5 (0.3–0.7) | 1.2 (0.7–1.6) |

Figures in parentheses are ranges.
[a] Due to tumor resection in 1 case.
[b] Due to irradiation technique not applicable in 2 patients.

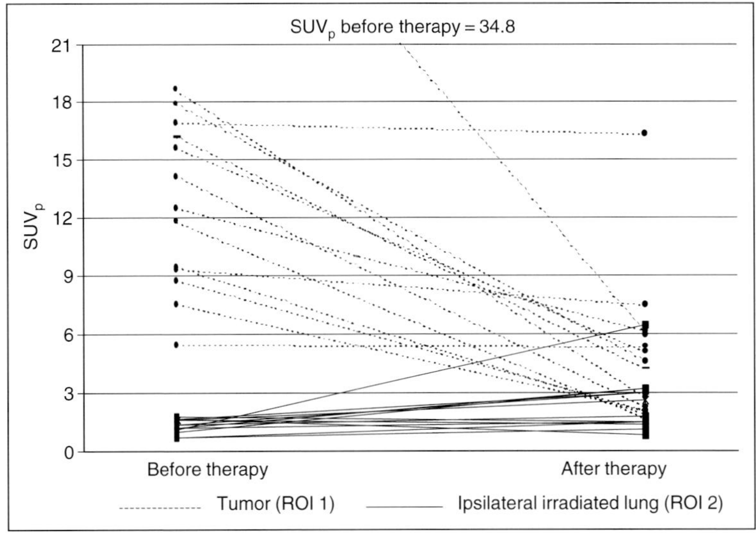


***Fig. 2.*** SUV$_p$ for tumor and ipsilateral irradiated lung of all patients individually before and after therapy.

After therapy, the SUVs in the unirradiated reference lung remained constant (ROI 4; table 1). In contrast, the median SUV$_a$ in the irradiated lung increased to 0.9 (ROI 2), and 0.7 (ROI 3), the median SUV$_p$ to 1.8 (ROI 2) and 1.3 (ROI 3), respectively (table 1, fig. 2).

***Table 2.*** Correlation of $SUV_a$ elevation and signs of pneumonitis in CT 3 months after irradiation

| | ROI 2 (ipsilateral lung) | | ROI 3 (contralateral lung) | | ROI 2 + ROI 3 | |
|---|---|---|---|---|---|---|
| | PET positive[a] | PET negative[b] | PET positive[a] | PET negative[b] | PET positive[a] | PET negative[b] |
| CT positive[c] | n = 11 (1.4) | n = 1 (0.6) | 0 | n = 4 (0.6) | n = 11 (1.4) | n = 5 (0.6) |
| CT negative[d] | n = 1 (0.7) | n = 2 (0.5) | n = 6 (0.8) | n = 3 (0.5) | n = 7 (0.8) | n = 5 (0.5) |

Figures in parentheses are mean $SUV_a$.
[a] $SUV_a > 0.7$.
[b] $SUV_a \leq 0.7$.
[c] Pneumonitis score $> 0$.
[d] Pneumonitis score $= 0$.

In comparison with the $SUV_a$ of the reference lung (ROI 4), elevated values, i.e. $SUV_a > 0.7$, were determined in 12/15 cases in ROI 2, and in 6/13 cases in ROI 3. The overall proportion of elevated $SUV_a$ in ROIs 1 and 2 was 18/28. In 3 patients, $SUV_p$ values in the ipsilateral irradiated lung exceeded a value of 3 reaching a range usually considered characteristic for malignant growth. The highest values ($SUV_p = 6.5$; $SUV_a = 3.0$) were determined in a patient whose tumor had been resected before radiotherapy, and who did not develop local recurrence until she died of brain metastases more than 18 months after treatment.

In 12 of the 15 patients, signs of radiation-induced pneumonitis were detected in ROI 2 of the restaging CT. Five of these cases showed slight, 4 patchy, and 3 dense impairment of transparency. In 9 of 13 cases, in ROI 3 no changes were observed, in 3 cases slight changes, and in 1 case patchy changes were recorded. The total proportion of CT changes in ROIs 2 and 3 was 16/28.

Signs of pneumonitis in CT and PET coincided in 13/15 cases in ROI 2 (table 2). In ROI 3 elevated $SUV_a$ without simultaneous CT changes were determined in 6/13 cases, while CT changes combined with normal $SUV_a$ were seen in 4/13 cases.

The severity of CT changes correlated significantly with the level of $SUV_a$ and $SUV_p$ ($r = 0.63$; $p = 0.01$; $SUV_a$/ipsilateral lung; fig. 3). In all 3 patients with $SUV_a$ and/or $SUV_p$ above 3, CT changes were present (fig. 3).

## Discussion

In the present study, CT and PET findings in irradiated lung tissue were compared in 15 patients treated for advanced NSCLC. Three months after

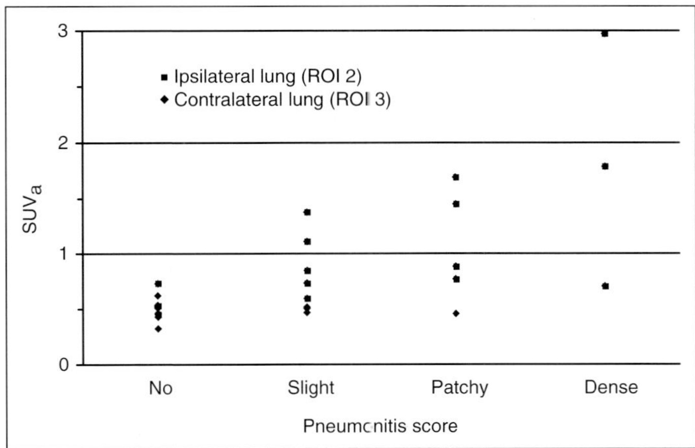


*Fig. 3.* Correlation of SUV$_a$ elevation and signs of pneumonitis in CT 3 months after irradiation.

irradiation, signs of pneumonitis were observed in 16/28 ROIs defined for irradiated lung tissue in the CT scans. This is in line with a previous report showing radiologic signs of pneumonitis in 60% of the patients 3 months after irradiation with a similar dose and fractionation schedule [13]. An increased FDG uptake was determined in 18/28 ROIs, suggesting that the incidence of pneumonitis determined 3 months after irradiation by PET is not or only marginally higher than by CT.

In most patients, only a slight elevation of the FDG uptake was observed. However, in 3 cases SUV$_p$ values were higher than 3, i.e. in a range usually considered characteristic of malignant growth. Sporadic cases of pronounced SUV elevations in patients with radiation pneumonitis have also been described by others [2, 5]. These findings support the need for concurrent CT studies when PET is used for restaging of lung cancer after radiotherapy. In the present investigation, CT changes typical for pneumonitis were present in all 3 patients with SUV values above 3.

Interestingly, a discordance was found between changes in PET (SUV$_a$ > the upper reference value of 0.7 in unirradiated lung tissue) and CT in 2/15 ROIs in the irradiated ipsilateral lung and in 10/13 ROIs in the irradiated contralateral lung. Due to the small number of patients included in the study, this observation may well be caused by statistical uncertainties or by the choice of the cut-off level for SUV$_a$. However, the results might also reflect differences in the sensitivity of CT and PET to detect specific components of the pathophysiology underlying radiation-induced pneumopathy, e.g. accumulation and activation of inflammatory cells versus edema. If so, [18]FDG-PET might become a useful, complementary tool for noninvasive studies on the mechanisms and the time course of radiation-induced pneumopathy in humans.

To the knowledge of the authors, no experimental studies have addressed the question as to which cell population(s) accumulate FDG in lung tissue after irradiation. However, several groups studied this question in non-radiation-induced inflammatory lung disease using microautoradiography. In a study on *Streptococcus pneumoniae*- and bleomycin-induced pneumonitis in rabbits, Jones et al. [9] showed that postmigratory neutrophils were the target cells of FDG accumulation. In bacterial infections in rats, Sugawara et al. [16] observed the highest FDG accumulation in inflammatory areas characterized by a high ratio between macrophages and polymorphonuclear leucocytes. From these studies it appears likely that the FDG-uptake might target macrophages and/or granulocytes also in radiation-induced pneumonitis. Specific studies are needed to verify this hypothesis.

## Conclusion

Three months after radiochemotherapy for lung cancer, an increased uptake of FDG could be demonstrated in 18/28 ROIs in irradiated lungs. While on average the level of the FDG enhancement measured by SUVs correlated with the severity of changes typical for radiation pneumonitis in CT, an interesting disparity of slight to moderate SUV enhancement in the PET study and changes in CT was found in individual patients. It is hypothesized that this finding reflects differences in the sensitivity of both methods to detect specific components of the pathophysiology underlying radiation-induced pneumopathy. If confirmed in further studies, [18]FDG-PET might become a useful, complementary tool for noninvasive studies on the mechanisms and the time course of radiation-induced pneumopathy in humans.

## References


1   Baumann M, Appold S, Geyer P, Knorr A, Voigtmann L, Herrmann T: Lack of effect of small high-dose volumes on the dose-response relationship for the development of fibrosis in distant parts of the ipsilateral lung in mini-pigs. Int J Radiat Biol 2000;76:477–485.
2   Cremerius U, Fabry U, Kroll U, et al: Clinical value of FDG PET for therapy monitoring of malignant lymphoma – results of a retrospective study in 72 patients. Nuklearmedizin 1999;38:24–30.
3   Dörr W, Baumann M, Herrmann T: Radiation-induced lung damage: A challenge for radiation biology, experimental and clinical radiotherapy. Int J Radiat Biol 2000;76:443–446.
4   Dwamena BA, Sonnad SS, Angobaldo JO, Wahl RL: Metastases from non-small cell lung cancer: Mediastinal staging in the 1990s – meta-analytic comparison of PET and CT. Radiology 1999;213:530–536.

5   Frank A, Lefkowitz D, Jaeger S, et al: Decision logic for retreatment of asymptomatic lung cancer recurrence based on positron emission tomography findings. Int J Radiat Oncol Biol Phys 1995;32:1495–1512.

6   Gambhir SS, Shepherd JE, Shah BD: Analytical decision model for the cost-effective management of solitary pulmonary nodules. J Clin Oncol 1998;16:2113–2125.

7   Hudson HM, Larkin RS: Accelerated image reconstruction using ordered subsets of projection data. IEEE Trans Med Imaging 1994;13:601–609.

8   International Commission on Radiation Units and Measurements (ICRU) Report 50: Prescribing, Recording, and Reporting Photon beam therapy, Bethesda ICRU 1993.

9   Jones HA, Clark RJ, Rhodes CG, Schofield JB, Krausz T, Haslett C: In vivo measurement of neutrophil activity in experimental lung inflammation. Am J Respir Crit Care Med 1994;149:1635–1639.

10  Kataoka M, Kawamura M, Itoh H, Hamamoto K: Ga-67 citrate scintigraphy for the early detection of radiation pneumonitis. Clin Nucl Med 1992;17:27–31.

11  McDonald S, Rubin P, Phillips T, Marks L: Injury to the lung from cancer therapy: Clinical syndromes, measurable endpoints, and potential scoring systems. Int J Radiat Oncol Biol Phys 1995;31:1187–1203.

12  Movsas B, Raffin TA, Epstein AH, Link CJ Jr: Pulmonary radiation injury. Chest 1997;111:1061–1076.

13  Nestle U, Nieder C, Walter K, et al: A palliative accelerated irradiation regimen for advanced non-small-cell lung cancer vs. conventionally fractionated 60 Gy: Results of a randomized equivalence study. Int J Radiat Oncol Biol Phys 2000;48:95–103.

14  Pieterman RM, van Putten JW, Meuzelaar JJ, et al: Preoperative staging of non-small-cell lung cancer with positron-emission tomography. N Engl J Med 2000;343:254–261.

15  Strauss LG, Conti PS: The applications of PET in clinical oncology. J Nucl Med 1991;32:623–648.

16  Sugawara Y, Gutowski TD, Fisher SJ, Brown RS, Wahl RL: Uptake of positron emission tomography tracers in experimental bacterial infections: A comparative biodistribution study of radiolabeled FDG, thymidine, $L$-methionine, 67Ga-citrate, and $^{125}$I-HSA. Eur J Nucl Med 1999;26:333–341.

17  Theuws JC, Kwa SL, Wagenaar AC, et al: Dose-effect relations for early local pulmonary injury after irradiation for malignant lymphoma and breast cancer. Radiother Oncol 1998;48:33–43.

18  Ukena D, Hellwig D, Palm I, et al: Value of positron emission tomography with 18-fluoro-deoxyglucose (FDG-PET) in diagnosis of recurrent bronchial carcinoma. Pneumologie 2000;54:49–53.

19  Vansteenkiste JF, Stroobants SG, De Leyn PR, Dupont PJ, Verbeken EK: Potential use of FDG-PET scan after induction chemotherapy in surgically staged IIIa-N2 non-small-cell lung cancer: A prospective pilot study. The Leuven Lung Cancer Group. Ann Oncol 1998;9:1193–1198.

20  Vansteenkiste JF, Stroobants SG, De Leyn PR, et al: Mediastinal lymph node staging with FDG-PET scan in patients with potentially operable non-small-cell lung cancer: A prospective analysis of 50 cases. Leuven Lung Cancer Group. Chest 1997;112:1480–1486.

21  LENT SOMA tables. Radiother Oncol 1995;35:17–60.



Dr. Ursula Nestle, Abteilung für Nuklearmedizin,
Radiologische Universitätsklinik, D–66421 Homburg/Saar
Tel. +49 6841 1624661, Fax +49 6841 1624861, E-Mail raunes@med-rz.uni-sb.de

Dörr W, Engenhart-Cabillic R, Zimmermann JS (eds): Normal Tissue Reactions in Radiotherapy and Oncology. Front Radiat Ther Oncol. Basel, Karger, 2002, vol 37, pp 34–37

# Efficacy and Quality of Life Outcomes of Epoetin-α in a Double-Blind, Placebo-Controlled, Multicentre Study of Cancer Patients Receiving Non-Platinum-Containing Chemotherapy

*T. J. Littlewood*

Department of Haematology, Oxford Radcliffe Hospital, Oxford, UK

Anaemia is a very common problem in patients with malignant disease. In the untreated patient it is most commonly referred to as 'anaemia of chronic disease'. The pathophysiology of this is not completely understood, but an increased production of interleukin 1 and tumour necrosis factor causing suppression of erythropoiesis and a relative erythropoietin deficiency certainly make important contributions [1]. Other frequent causes of anaemia include blood loss, haematinic deficiency, marrow infiltration by tumours and haemolysis. Chemotherapy and radiotherapy will exacerbate the severity of anaemia.

Elegant work from Miller et al. [9] showed that although anaemic patients with cancer have higher than normal levels of erythropoietin, the increase in erythropoietin levels is significantly less than that seen in patients with comparably severe anaemia due to iron deficiency or haemolysis. Typical symptoms of anaemia include fatigue [11], lethargy, breathlessness and swollen feet and ankles.

Several studies have been performed which investigated the therapeutic benefit of erythropoietin in anaemic patients with cancer. The results from non-randomized and small randomized trials showed that 30–80% of patients responded to erythropoietin with a haemoglobin increase of more than 2 g/dl and most studies showed a response rate of around 50% [2, 8, 10]. These early studies also showed a reduction of approximately 50% in transfusion requirement in treated patients. Subsequently, two very large community-based trials were reported from the USA [4, 5]. Both studies were designed to identify the impact of treatment with recombinant erythropoietin (epoetin-α) on haemoglobin levels in patients with cancer. The first of the studies retrospectively looked at quality of life issues, whereas the second study took a prospective view of the quality of life and correlated the changes with the anti-tumour response. The conclusions from these two studies are that epoetin-α improves the quality of life in these patients, that the improvement in quality of life correlated with the improvement in haemoglobin up to a level of greater than 12.0 g/dl [3] and that it occurred independently of the anti-tumour response. In order to confirm the results of the small randomized studies and the data from the large non-randomized community studies, a randomized, double-blind, placebo-controlled, multicentre study was set up to assess the effect of treatment with epoetin-α in anaemic cancer patients receiving non-platinum-containing chemotherapy [6].

## Material and Methods

This double-blind placebo-controlled trial involved 375 patients at 73 sites in 15 countries, all in Europe apart from 1 study centre in South Africa. Recruitment began in 1996. The patients were all older than 18 years, had either a solid tumour or a non-myeloid haematological malignancy and were expected to live more than 6 months. The patients were matched for age, sex and underlying tumour type between the treatment and placebo groups. Approximately 54% of the patients had a solid tumour and 46% a haematological malignancy. All patients had a haemoglobin level that was either less than or equal to 10.5 g/dl or that lay between 10.5 and 12.0 g/dl and had been subject to a recent drop of more than 1.5 g/dl. Patients were administered either epoetin-α or placebo, in a two-to-one ratio in favour of epoetin-α, three times a week subcutaneously for a maximum of 28 weeks, according to a treatment protocol which took into account variations in the patient's haematological status during the study period. Transfusions were permitted, but only if the haemoglobin level dropped below 8 g/dl.

The study was restricted to patients receiving chemotherapy which excluded platinum to expand the proven efficacy of epoetin-α in anaemic patients receiving platinum-based chemotherapy.

## Results

As in earlier studies, epoetin-α was found to be well tolerated with no difference in the frequency of side effects between the treatment and placebo groups. From baseline to study end the mean change in the haemoglobin in the epoetin-α-treated group was 2.2 g/dl contrasting with a rise of 0.5 g/dl in the placebo-treated group (p < 0.001).

From the end of week 4 to the end of the study, 24.7% (62/251) of epoetin-α-treated patients and 39.5% (49/124) of placebo-treated patients were transfused (p = 0.0057). The difference in transfusion need occurred irrespectively of whether the patients had a haematological or a solid tumour.

Three quality of life scales were used in the study; the visual linear analogue scale, the Functional Assessment of Cancer Therapy (FACT) scale and the short form 36. Epoetin-$\alpha$ patients scored significantly better on all three linear analogue scales (energy $p < 0.001$; activities $p < 0.01$, and overall quality of life $p < 0.01$) than placebo-treated patients.

Similar statistically significant improvements were seen in the epoetin-$\alpha$-treated patients based on the FACT-General scale ($p < 0.05$), the FACT-Fatigue scale ($p < 0.01$) as well as on the FACT-Anaemia scale ($p < 0.01$). Using the short form 36, a trend to improvement was seen in the epoetin-$\alpha$-treated patients, but this did not quite achieve statistical significance.

Before this study was unblinded, a further analysis of the results [7] suggested that there may be a survival advantage for patients treated with epoetin-$\alpha$ compared to placebo-treated patients, but this finding requires confirmation.

## Conclusions

Epoetin-$\alpha$ is a safe treatment which increases the haemoglobin concentration in the majority of anaemic patients with cancer. This increase results in a reduction in transfusion need and, perhaps most significantly, in an overall improvement in the patients' quality of life.

## References


1   Cazzola M, Mercuriali F, Brugnara C: Use of recombinant erythropoietin outside the setting of uremia. Blood 1997;89:4248–4267.
2   Cazzola M, Messinger D, Battistel V, et al: Recombinant human erythropoietin in the anemia associated with multiple myeloma or non-Hodgkin's lymphoma: Dose finding and identification of the predictors of response. Blood 1995;86:4446–4453.
3   Cleeland CS, Demetri GD, Glaspy J, Cella D: Identifying hemoglobin level for optimal quality of life: Results of an incremental analysis. Proc ASCO 1999;18:574a.
4   Demetri GD, Kris M, Wade J, Degos L, Cella D: Quality of life benefit in chemotherapy patients treated with epoetin alpha is independent of disease response or tumour type: Results from a prospective community oncology study. J Clin Oncol 1998;16:3412–3425.
5   Glaspy J, Bukowski R, Steinberg D, Taylor C, Tchekmedyian S, Vadhan-Raj S: Impact of therapy with epoetin alpha on clinical outcomes in patients with nonmyeloid malignancies during cancer chemotherapy in community oncology practice. Procrit Study Group. J Clin Oncol 1997;15:1218–1234.
6   Littlewood TJ, Bajetta E, Cella D: Efficacy and quality of life outcomes of epoetin alfa in a double-blind, placebo-controlled multicenter study of cancer patients receiving non-platinum containing chemotherapy. Proc ASCO 1999;18:574a.
7   Littlewood TJ, Rapoport B, Bajetta E, Nortier JWR: Possible relationship of hemoglobin levels with survival in anemic cancer patients receiving chemotherapy. Proc ASCO 2000;19:605a.

8 Ludwig H, Fritz E, Kotzmann H, Hocker P, Gisslinger H, Barnas U: Erythropoietin treatment of anemia associated with multiple myeloma. N Engl J Med 1990;322:1693–1699.

9 Miller CB, Jones RJ, Piantadosi S, Abeloff MD, Spivak JL: Decreased erythropoietin response in patients with the anemia of cancer. N Engl J Med 1990;322:1689–1692.

10 Osterborg A, Boogaerts MA, Cimino R, et al: Recombinant human erythropoietin in transfusion-dependent anemic patients with multiple myeloma and non-Hodgkin's lymphoma – A randomized multicentre study. Blood 1996;87:2675–2682.

11 Vogelzang NJ, Breitbart W, Cella D, et al: Patient, caregiver, and oncologists perceptions of cancer-related fatigue: Results of a tripart assessment survey. Semin Hematol 1997;34(suppl 3):4–12.

T.J. Littlewood, MB, BCh, FRCP, FRCPath, MD, Department of Haematology,
Oxford Radcliffe Hospital, Oxford OX3-9DU (UK)
Tel. +44 186 522 0364, Fax +44 186 522 1778, E-Mail tim.littlewood@orh.nhs.uk

Dörr W, Engenhart-Cabillic R, Zimmermann JS (eds): Normal Tissue Reactions in Radiotherapy and Oncology. Front Radiat Ther Oncol. Basel, Karger, 2002, vol 37, pp 38–42

# Physical and Psychosocial Rehabilitation of Cancer Patients

*Eva-Susanne Strobel*

Paracelsus-Klinik am Schillergarten, Bad Elster, Deutschland

The diagnosis of cancer is a psychologic trauma for patients and their families. Commonly, cancer is perceived as a debilitating disease leading to pain, misery and death. The diagnosis of cancer is often associated with emotional distress, anxiety, hopelessness and depression. Psychosocial intervention has positive effects on psychic well-being and on quality of life.

In addition to psychosocial problems caused by cancer, there are physical deficits resulting from the disease itself and from its therapy. Examples of disease-associated impairment are hemiplegia caused by brain tumors, loss of vision caused by retinoblastomas, hoarseness caused by lung cancers, constipation caused by colon cancers and ascites caused by ovarian cancers. In addition, therapy-associated problems can occur after surgery, radiotherapy and chemotherapy. Dyspnea after pneumonectomy, nutritional problems after gastrectomy for stomach cancer, pain and swelling of the arm and decrease of motion in the shoulder joint after mastectomy and lymphadenectomy, or decreased mobility as a result of amputations of extremities for sarcomas are attributed to surgery. Radiotherapy can cause organ-specific side effects, like oral mucositis, pneumonitis, pericarditis, enteritis, radiomyelitis, nephritis or xerostomia. Chemotherapy can result in leukopenia, infection, thrombocytopenia, bleeding, anemia, fatigue, polyneuropathy, renal insufficiency, pulmonary fibrosis and cardiomyopathy. Both radio- and chemotherapy can induce infertility and secondary malignancies.

Rehabilitation should restore the patient's condition as much as possible [2]. The aims of oncologic rehabilitation can be split into somatic, functional, psychosocial and educative aims. Somatic aims include pain control, improvement of the range of motion, improvement of pulmonary function, reduction of climacteric symptoms and reduction of lymphedema. Functional aims are the compensation for limitations through exercising the remaining functions. Examples are training of the pelvic floor muscles in addition to electrotherapy for urinary incontinence or training of esophageal speech after laryngectomy [3]. Psychosocial aims include optimal restoration of health, maximum potential for normal living, improvement in quality of life, physical and emotional fitness, improvement in coping, reintegration into daily living and family life, preservation or restoration of the working capacity, reintegration into professional and social life, information about self-help groups and further care at home. Educative aims can be information about the disease, management of stomas as well as prevention and treatment of lymphedema.

To respond to the physical, spiritual and emotional needs of the patients, a team approach is required. Medical doctors, psychologists, physical therapists, dieticians, social workers, art and occupational therapists and nurses must belong to this rehabilitation team. The patients' physical impairment and psychosocial problems are analyzed on admission. Mobility, range of motion, strength and endurance are evaluated and deficits are defined; lymphedema is quantified. If necessary, questionnaires can be used for the evaluation of depression, anxiety and pain. Further diagnostic tests include laboratory examinations with complete blood counts, erythrocyte sedimentation rate, blood chemistry and tumor markers. If required, electrocardiography, exercise electrocardiography, ambulatory 24-hour blood pressure and electrocardiography monitoring, echocardiography, sonography, pulmonary function tests, radiologic examinations and examinations by various consultants are performed.

Physical rehabilitation includes physiotherapy, which can be done alone or in a group. In the case of tense muscles it is preferably performed in warm water where the force of gravity is diminished. Lymphedema is treated with manual lymphatic drainage and, if necessary, with compression bandages, elastic stockings and afterwards with physiotherapy. Massages are given in combination with thermotherapy, which can be applied warm or cold. Electrotherapy and ultrasonic therapy can be added. Therapeutic exercises, ergometer training and swimming are used to train specific muscles for the compensation of deficits and to increase the range of motion, physical fitness and endurance.

Psychosocial rehabilitation comprises information about the disease, treatment and prognosis. Therefore special training courses are offered for the prevention of lymphedema, management of breast prostheses, stomas or of the tracheostomy tube. Seminars are performed on various cancer types, prevention of stress, physical training, healthy cooking, low-cholesterol and low-fat diet, hypertension and diabetes. Relaxation techniques such as progressive muscle relaxation, autogenic training and biofeedback are offered.

Psychologists offer individual therapy for specific problems. Also visualization according to Simonton and group therapy for the management of anxiety, depression and pain are provided. For example, in pain management strategies, the presence of pain is analyzed and the cause treated, if possible. Pain control strategies can be learnt and trained using a combination of physical therapy, psychologic strategies and medication. Physical therapy can include physiotherapy, thermo-, electro- and ultrasonic therapy. Psychologic strategies such as relaxation and attention focussing can add to the effect of physical therapy. Finally, different pharmacologic classes of analgesics may be administered.

In painting groups, the patients develop their own creative potential under the guidance of the art therapist, and the paintings are subsequently discussed. Occupational therapy offers silk painting, patchwork, sewing, painting of ceramics, dancing, excursions, walking, hiking and slide shows.

Social interventions include reintegration into the professional and social life and information about benefits for handicapped people. Further education, occupational training and job placement may be important issues for patients unable to work in their previous jobs due to their disabilities. Questions regarding disability pensions are discussed if a return to any job is impossible. Self-help groups provide support by people involved in the cancer problem. Information about the possibilities of further care at home is given.

Information, relaxation, psychotherapy and social interventions together help the patient to develop coping strategies [1]. All these therapies are given in a supportive and understanding atmosphere where the patient can trust his therapists.

## Example

As an example, the physical and psychosocial rehabilitation of a 52-year-old woman with cancer of the right breast, $pT_{1c}\ pN_0\ M_0\ G2\ ER+\ PR+$, is reported. The patient had undergone breast-conserving surgery and subsequent radiotherapy; tamoxifen was administered daily. Apart from her malignancy, she was healthy and had worked as a secretary until diagnosis. The physical deficits include discrete lymphedema in the right arm and limited range of motion in the right shoulder joint. The patient is afraid of a recurrence and suffers from somnipathy. She frequently wakes up, dreaming of her aunt who has recently died from metastatic breast cancer. Hypercholesterinemia was detected by laboratory analysis.

The physician discussed with the patient which aims of rehabilitation she wants to reach during the time period of 3 weeks: the lymphedema should be reduced and the range of motion improved. Coping with the disease should be improved and thus the sleeping problems alleviated.

***Table 1.*** Weekly treatment plan for a 52-year-old woman with breast cancer $pT_{1c}$ $pN_0$ $M_0$

|  | Monday | Tuesday | Wednesday | Thursday | Friday |
|---|---|---|---|---|---|
| 7.00 a.m. | morning gymnastics | morning gymnastics | morning gymnastics | morning gymnastics | morning gymnastics |
| 8.00 a.m. | breakfast | breakfast | breakfast | breakfast | breakfast |
| 9.00 a.m. | manual lymphatic drainage | group physiotherapy | manual lymphatic drainage | group physiotherapy | manual lymphatic drainage |
| 10.00 a.m. | physiotherapy |  | physiotherapy |  | physiotherapy |
| 11.00 a.m. |  | physician |  | psychologist |  |
| 12.00 noon | lunch | lunch | lunch | lunch | lunch |
| 1.00 p.m. | progressive relaxation | visualizing group | progressive relaxation | visualizing group | progressive relaxation |
| 2.00 p.m. |  | art group: painting | cooking class | art group: painting |  |
| 3.00 p.m. | training courses[1] | art group: painting | cooking class | art group: painting | training courses[1] |
| 4.00 p.m. |  |  |  |  |  |
| 5.00 p.m. | swimming | swimming | swimming | swimming | swimming |
| 6.00 p.m. | dinner | dinner | dinner | dinner | dinner |
| 7.00 p.m. | occupational therapy |  |  |  | occupational therapy |

[1] The training courses give information and allow discussion about the following subjects: breast cancer, management of breast prostheses, prevention of lymphedema, prevention of stress, physical training, low-cholesterol and low-fat diet.

The treatment plan (table 1) was designed together with the patient. An opportunity to talk to other patients and to discuss her problems with the physician and the psychologists was established in order to assist in the development of coping strategies. At least weekly the progress is evaluated. Before discharge she discussed her rehabilitation results and improvements with her physician: she felt much better, there was no lymphedema left and the mobility of the right shoulder had improved significantly. She now sleeps well without any further nightmares, is motivated to continue physiotherapy for the right shoulder at home and will join a self-help group. She plans to go back to work after another few weeks.

## Conclusions

In conclusion, the combination of physical therapy, medical and psychologic help, information and training in the management of cancer, relaxation, art and occupational therapy enables the patients to cope with cancer. It gives the patients new energy and strength, and help to reintegrate them into their social and professional life.

## References

1   Joly F, Henry-Amar M, Arveux P, et al: Late psychosocial sequelae in Hodgkin's disease survivors: A French population-based case-control study. J Clin Oncol 1996;14:2444–2453.
2   Razavi D, Delvaux N: The psychiatrist's perspective on quality of life and quality of care in oncology: Concepts, symptom management, communication issues. Eur J Cancer 1995;31A(suppl 6): 25–29.
3   Schraub S, Bontemps P, Mercier M, Barthod L, Fournier J: Surveillance and rehabilitation of cancer of upper respiratory and digestive tracts. Rev Prat 1995;45:861–864.


Priv.-Doz. Dr. med. Eva-Susanne Strobel, Paracelsus-Klinik am Schillergarten,
Martin-Andersen-Nexö-Strasse 10, D–08645 Bad Elster (Germany)
Tel. +49 37437 70421, Fax +49 37437 70304, E-Mail bad_elster@paracelsus-kliniken.de

Dörr W, Engenhart-Cabillic R, Zimmermann JS (eds): Normal Tissue Reactions in Radiotherapy and Oncology. Front Radiat Ther Oncol. Basel, Karger, 2002, vol 37, pp 43–48

# Personal and Institutional Liability

*A. Resch-Holeczke[a], H. Ofner[b], R. Pötter[a]*

[a] Klinik für Strahlentherapie und Strahlenbiologie (Head: R. Pötter) und
[b] Institut für Rechtsvergleichung, Universität Wien, Österreich

## Survey

Various legal claims may result from the practice of medicine. Such claims have to be dealt with separately according to the branch of law concerned. Basically, a distinction has to be made between consequences under criminal and civil law.

## Criminal Law

Only in rare, serious cases is medical malpractice followed by consequences under criminal law. The main reasons for this are to be found in the motivation of the patient who is less interested in the punishment of the physician than in financial compensation. As a result, in many cases the patient does not file a report with the police. However, the fact that the costs of medical expert opinions in criminal proceedings are to be borne by the state and not by the party losing the proceedings, as in civil proceedings, and that these records may be used in subsequent civil proceedings, constitutes a certain incentive for criminal prosecution.

If criminal proceedings are nevertheless instituted, in most cases the offence sued for is negligent bodily injury; in Austria the offence of unlawful therapeutic treatment is also possible.

## Civil Law

*Theory of Liability*

According to German and Austrian law, claims for damages may only be asserted if the following elements have been established.

- The patient must have suffered damage.
- The physician's conduct causing the damage must be contrary to law. The unlawfulness may result from a violation of the treatment contract on the one hand, or from a violation of so-called absolute rights (e.g. bodily integrity) of the patient on the other hand.
- There must be a causal connection between the doctor's unlawful conduct and the harm complained about (causation).
- The damage must have been inflicted by culpable negligent conduct of the physician.

*Principle of Culpable Negligence*

In contrast to several other European legal systems, German and Austrian law thus starts from the principle of culpable negligence. Therefore, a claim for damages is only possible if the damage was inflicted negligently by the attending physician – by his objective failure to exercise due care.

Amount and assessment of damages are mainly dependent on the degree of negligence. Basically, a distinction is made between intentional and negligent conduct. A person acts intentionally if he deliberately causes a harmful result, or if he seriously contemplates and accepts this result. He acts negligently if he fails to exercise the care he ought to exercise under the given circumstances or is capable of exercising according to his mental and physical abilities.

*Damages to Be Compensated*

German and Austrian law of damages basically starts from the principle of restitution in kind. That means that the patient's previous situation has to be restored as if the physician's conduct causing the damage had not taken place.

Therefore, costs to be compensated are costs of medical treatment which would not have been necessary with a careful diagnosis and corresponding therapy, funeral costs in case of the patient's death, compensation for loss of earnings which would not have been incurred in case of a careful diagnosis and corresponding therapy, financial compensation for feelings of pain suffered – 'nonpecuniary damages' – which could have been avoided with a careful diagnosis and corresponding therapy, and claims for maintenance of the patient's relatives which the latter is no longer able to fulfill due to the damage caused by the physician.

*Bases for Claims*

According to German and Austrian law, liability of the physician under civil law arises from two different bases for claims requiring different legal elements of the offence.

*Violation of the Treatment Contract.* The basis of the physician-patient relationship under civil law is the treatment contract. This contract can also be concluded by mutual consent. Contractual obligations, the contents and scope of which are basically dependent on the agreement concluded, arise from the treatment contract for the respective contracting party of the patient. The primary contractual obligation is the performance of the agreed medical service (treatment and/or diagnosis). In addition, however, various duties to protect and exercise due care arise for the patient's contracting partner. The same is thus also obliged to inform the patient about the detailed circumstances of his individual case, in particular, on the treatment or diagnostic measure to be recommended. If the contracting partner violates any of the obligations arising from the treatment contract, e.g. the contractual duty to inform, the patient is entitled to assert a claim for damages suffered as a result of insufficient information supplied by the contracting partner. In case the treatment contract was not concluded with the physician but with the hospital institution, the contractual liability is to be assumed for the hospital institution. However, the latter may have claims against the physician employed.

*Tort Liability.* Tort liability is not bound to an existing contractual relationship but results from the performance of the diagnostic measure or curative treatment which leads to an interference with the bodily integrity of the patient. Tort liability may therefore not only affect physicians in private practice and chief doctors but also trainee, assistant and senior doctors in hospitals. Concurring claims for torts and contractual damages may be asserted. Thus, e.g. the patient may take action against the hospital institution on the basis of the treatment contract and against the attending hospital doctor on the basis of the tort.

*Burden of Proof.* In tort liability, the burden of proof of existing damage as well as unlawfulness, causal relationship and culpable negligence committed by the physician lies with the patient. In case of contractual liability, however, the patient is only required to prove damage, unlawfulness and causal relationship, while the physician's culpable negligence is assumed even without concrete evidence. If there is no negligence, the doctor is forced to prove his innocence.


*Malpractice*

Incorrect treatment has been performed if the medical diagnosis or therapy measure was not taken or judged lege artis.

In radiation therapy, malpractice may be due to various types of incorrect conduct: use of insufficient diagnostic tools, failure to detect a tumor irrespective of the use of the required diagnostic tools, incorrect planning of irradiation fields, incorrect calculation of the radiation dose, and incorrect prescription of dose and/or fractionation.

However, since the Austrian law of damages starts from the principle of negligence, a claim for damages by the patient may be asserted only if the types of incorrect conduct stated above are caused by careless conduct of the physician. Therefore, from the perspective of the ordinary skilled radiotherapist, it is to be examined whether the same would have committed the said malpractice. In particular, the questions which diagnostic measures are to be taken, how planning has to be conducted and which dose in which fractionation is to be administered are determined by the current standard of medical science.

To restrict the liability risk, it therefore seems advisable for medical professional societies to establish guidelines on the standard of diagnostic measures.

### Duty to Inform

The duty to inform the patient arises from the treatment contract. Noninformation or insufficient information supplied to the patient constitutes a contractual violation. In the field of tort liability, information given to the patient also plays an important role, since a therapeutic measure interfering with the bodily integrity of the patient excludes liability only if the patient gave his or her consent. However, this consent is to be based on comprehensive information supplied to the patient. The attending physician is thus obliged to relate to the patient any circumstances concerning the treatment to enable the patient to make a decision independently.

Scope and Contents of Information

The scope of the information supplied depends on the kind of treatment administered. Basically, anything necessary for the patient's decision is to be related to the patient. In this respect, legal precedents have developed two essential guidelines. On the one hand, the scope of information required depends on the potential risks and, on the other hand, on the necessity of the treatment. The more serious the potential consequences of the diagnostic or therapeutic measure for the patient, and the weaker the indication of the diagnostic or therapeutic measure, the greater the physician's duty of disclosure.

Looking at the possible types of information, three groups of information contents may be drawn up.

*Risk Information.* The physician is obliged to inform the patient on the kind and probability of typical risks involved in the treatment. The typicality is not necessarily dependent on the probability of its occurrence.

*Course Information.* Further, the physician is obliged to inform the patient about the medical report, kind, scope, urgency, probable course and necessary consequences of the planned diagnostic or therapeutic measure. Information on the course also includes the discussion of the consequences to be expected in case of nontreatment.

*Alternative Treatment.* The patient must be informed about possible alternative therapies and their benefits and disadvantages.

Place and Time of Information
Information must be provided at a time when the patient is still able to make a decision independently. This means that the information must be supplied in time so that the patient does not consider him- or herself forced to consent as a result of accompanying circumstances. In any case, it is too late to give information on operation risks in the anteroom.

Kind of Information
The kind of information supplied depends on the purpose of the information. As stated above, the purpose of the information is to disclose any facts relevant for the patient to make an independent decision for or against the diagnostic or therapeutic measure. Therefore, the information, as the basis of the decision to be taken, has to be supplied in a way appropriate for the patient's level of understanding. The attending physician thus has to inform the patient in an understandable manner corresponding to the patient's personality, comprehension and education.

This requirement of information makes it necessary that there is an informative oral consultation during which the physician may get an impression of the patient's specific need for information. The use of printed forms may well reduce the liability risk. However, complete coverage can only be obtained in a personal conversation taking into consideration the personality of the patient. The informed consent form may be of good service as a written proof of scope, contents, time, place and kind of information supplied.

*Liability of the Hospital Institution*
If the treatment takes place in a hospital, as a rule, the treatment contract is not concluded between the patient and the attending physician but between the patient and the hospital institution. In this case, the attending doctor acts as the vicarious agent of the hospital without being a contracting partner of the patient. Only chief physicians become contractual partners of the patients, if the patient is covered by private supplementary health insurance. In this case, the hospital institution has to take over the vicarious liability for the physicians acting as vicarious agents of the hospital. On the one hand, the conduct giving rise to liability may result from malpractice or organizational failure. The latter may e.g. consist of the fact that the head of the organization in a department and/or

head of the hospital fails to provide sufficient staff or state-of-the-art equipment. Thus, a work schedule violating legal provisions on working time which causes malpractice of a physician due to his or her being overtired would be considered to be an organizational failure.

Exempt therefrom are patients covered by private supplementary health insurance. In their case, the treatment contract is not concluded with the hospital institution but between the patient and the chief physician. The patient and the hospital institution only enter into a contractual relationship relating to nursing care and accommodation.

Dr. A. Resch-Holeczke, Klinik für Strahlentherapie und Strahlenbiologie,
AKH Wien, Währinger Gürtel 18–20, A–1090 Wien (Austria)
Tel. +43 1 40 400 2692, Fax +43 1 40 400 2690, E-Mail alexandra.resch@univie.ac.at

Dörr W, Engenhart-Cabillic R, Zimmermann JS (eds): Normal Tissue Reactions in Radiotherapy
and Oncology. Front Radiat Ther Oncol. Basel, Karger, 2002, vol 37, pp 49–56

# Radiotherapy-Related Fatigue and Exercise for Cancer Patients: A Review of the Literature and Suggestions for Future Research

*Fernando Dimeo*

Institute of Sports Medicine, Freie Universität Berlin, Germany

Fatigue, defined as a feeling of weariness, tiredness, or lack of energy, is a well-documented phenomenon in cancer patients during chemo- and radiotherapy. It can affect all aspects of a person's life; however, little is known about the origin of this symptom or its prevalence and severity throughout the disease and treatment.

Several aetiological mechanisms have been postulated to explain the development of fatigue in cancer patients. These include psychosocial stress, pain, electrolyte and fluid disturbance, anaemia, poor nutritional status, weight loss, changes in the concentration of metabolically active molecules as a result of the interaction between the tumour and host defence system, intercurrent systemic pathology, drugs with action on the central nervous system and sleep disturbances [23, 27]. However, since no consistent relationship between these variables and fatigue has been found, the actual causes of impaired physical function in this setting are not yet fully understood.

When considered from a teleological point of view, fatigue is a normal and necessary instrument of physiological self-regulation. Fatigue that appears after intense or prolonged activities protects the body from exaggerated or harmful efforts. However, fatigue can also turn pathological when it appears during usual activities, persists for a long time, does not improve after rest, or becomes severe enough to force patients to reduce their level of activity.

The fact that fatigue has recently been recognized as one of the most frequent secondary effects of cancer treatment has given rise to considerable interest in the aetiology of fatigue and its prevalence in cancer patients.

## Fatigue in Cancer Patients Receiving Radiotherapy

Several studies have assessed the fatigue experienced by patients receiving radiotherapy. King et al. [16] evaluated the fatigue patterns in 79 cancer patients during radiation treatment. Patients were interviewed weekly during and monthly after the end of treatment. The prevalence of fatigue increased from 60% during the 1st week of treatment to 93% 2 weeks later and gradually decreased to 46% 3 months posttreatment.

In a large prospective study, Smets et al. [24] evaluated the prevalence and intensity of physical and mental fatigue in 250 cancer patients undergoing radiotherapy. About 40% of the patients reported having been tired most of the time during radiation. Fatigue still persisted after treatment in 50% of the patients. For most patients, fatigue was one of the three symptoms that caused the greatest distress. Finally, high fatigue scores were related to a more severe impairment in the patient's ability to perform daily activities.

In a further study, the same investigators assessed the prevalence and course of fatigue in disease-free patients after the conclusion of treatment [25]. A total of 154 patients were evaluated and their symptoms were compared with a reference group of healthy persons. The two groups showed no differences in global fatigue scores and in the domains of general fatigue, physical fatigue, reduced activity and reduced motivation. However, fatigue was stable in more patients than controls during the month before assessment, suggesting that it is a more chronic condition in patients.

Some studies suggest that fatigue may be more intense in patients undergoing radiotherapy for cancer treatment than in those patients receiving chemotherapy [23]. Berglund et al. [1] compared two groups of breast cancer patients (172 after radiation and 201 after chemotherapy) between 2 and 10 years after treatment; none of the patients had a relapse at the time of evaluation. Patients after radiotherapy reported loss of stamina more frequently than those who underwent chemotherapy.

Acute and chronic side effects of treatment seem to play an important role in the genesis of fatigue. In a randomized study, a low-fat, low-lactose diet resulted in a reduction of diarrhoea and fatigue and a higher functional status in women receiving pelvic irradiation for gynaecological malignancies [3]. In a prospective study, 75 patients with head and neck cancer were evaluated before the onset of radiotherapy and 6 and 12 months thereafter. During this time, there was a significant impairment of physical function and an increase of fatigue scores. In this study, the severity of fatigue was clearly related to a loss of functional status. However, despite severe physical deterioration during radiotherapy, emotional function improved significantly and depression scores did not increase, suggesting a lack of a relationship between fatigue and psychological distress [4]. These findings were reproduced in a further study including 65 patients with laryngeal cancer [5].

Other reports about the relationship between fatigue and mood disturbance have yielded contradictory findings. In a convenience sample of 24 patients receiving radiotherapy for bone metastases, fatigue was related to sleep disturbances and pain [18]. However, in a prospective study of fatigue in prostate cancer patients undergoing radiotherapy [21], there was no relationship between fatigue, depression and sleep disturbance.

Most observations suggest that fatigue during radiotherapy is related to an indirect effect of ionizing radiation on the body. Greenberg et al. [13] evaluated the effects of local irradiation in women with breast cancer. Fatigue increased during treatment and reached a plateau in the 4th week (after an average of 17 fractions), which was maintained until the end of treatment. No changes in depression scores were observed during treatment. Other markers (thyroid hormones, haematocrit, weight loss, cardiac function) showed no correlation with fatigue. Since irradiation was restricted to a small volume of the body and showed no correlation with depression, the authors concluded that fatigue associated with radiotherapy is due to a systemic reaction to tissue injury.

A further study yielded evidence that fatigue may be secondary to a decline in neuromuscular efficiency [22]. The authors carried out three evaluations of neuromuscular fatigue of the tibialis anterior muscle, cardiopulmonary fitness and psychological subjective fatigue in 13 prostate cancer patients: before irradiation, at the end of treatment, and 5–6 weeks after radiotherapy. A significant decline in neuromuscular efficiency was observed at the end of treatment. The phenomenon was attributed to an increased release of cytokines as a consequence of tissue necrosis after radiotherapy. However, since the concentration of cytokines in blood was not assessed, no objective data provided support for this hypothesis.

## Therapeutic Interventions in Radiotherapy-Related Fatigue

Though cancer patients often identify fatigue as a major problem, this symptom has been ignored in most oncology rehabilitation programmes. Furthermore, there are few data on fatigue patterns, exacerbating or relieving factors, aetiological mechanisms, the intensity of this symptom at different stages of disease and treatment, and its prevalence in various groups of cancer patients. This lack of information has made it difficult to develop therapies for cancer fatigue.

Graydon et al. [12] identified the fatigue-reducing strategies used by patients receiving cancer treatment. The main coping strategies were sleep and exercise, the latter being more effective in reducing fatigue. The authors mention that patients undergoing treatment for cancer are often advised to limit their activity and get plenty of rest. These may be effective strategies in acute situations of fatigue. However, in patients suffering from long-lasting fatigue, they can result in physical deconditioning and therefore in increased fatigue.

Two studies have evaluated the effects of psychotherapy on cancer patients with fatigue [10, 11]. In the first study, 24 patients receiving radiotherapy were randomly selected for 10 weeks of group psychotherapy, 90 min per week. Another 24 patients served as controls. Emotional and physical symptoms were evaluated at the beginning, in the middle, at the end of radiotherapy, and 4 and 8 weeks after treatment. Psychotherapy resulted in a slight reduction of emotional symptoms in cancer patients 4 weeks after treatment [10]. In a further study, the same investigators evaluated the effects of individual psychotherapy during radiation therapy [11]. Forty-eight patients undergoing radiotherapy were given weekly psychotherapy sessions for 10 weeks; another 52 patients served as controls. Both emotional and physical distress were reduced in both groups at the end of treatment. However, the reduction was significantly greater in the patients receiving psychotherapy than in the control group.

## Effects of Exercise on Fatigue and Physical Performance

In recent years, there has been a growing interest in the effects of exercise as therapy for cancer fatigue. Traditionally, exercise programmes in oncological rehabilitation have been limited to physical therapies addressing specific impairments caused, for example, by amputation or surgery. The concept of physical activity as a therapy for cancer fatigue has not yet been fully accepted. Many patients and physicians believe that vigorous exertion is potentially harmful, especially after recent exposure to cardiotoxic agents such as anthracyclines. Furthermore, prescribing physical activity to patients suffering from fatigue may appear counterintuitive.

However, adaptive changes generated by exercise may counteract several negative effects of treatment and disease on functional ability. Physical activity results in an increase in muscle mass and plasma volume, improved lung ventilation and perfusion, increased cardiac reserve, and a higher concentration of oxidative muscle enzymes. Moreover, resistance exercise has been shown to reduce the loss of muscle mass related to treatment with high-dose corticoids [2, 14] and some evidence suggests that exercise may reduce the cardiotoxic effects of anthracyclines [15]. Physical activity can therefore decrease fatigue by normalization of the physical performance.

One of the most promising interventions in this field is aerobic exercise, defined as the rhythmical contraction and relaxation of large muscle groups over a prolonged time. Aerobic exercise forms are, for example, walking, jogging, biking, rowing and swimming. Several studies have evaluated the effects of structured and nonstructured aerobic exercise programmes on the physical performance and well-being of cancer patients [17, 19, 26].

## Endurance Exercise as Treatment for Cancer Fatigue

In recent reports [6–8], we have described the effects of exercise programmes for patients with solid tumours and haematological neoplastic diseases during and after intensive treatment (high-dose chemotherapy, HDC). In the first study, 20 patients with haematological malignancies after bone marrow transplantation participated in a structured training programme consisting of daily walking on a treadmill [8]. Patients were enrolled in the study as they had a trilinear haematopoietic reconstitution and a stable clinical condition. Training was started 30 ± 6 days (ranging from 18 to 42 days) after bone marrow transplantation. The training programme consisted of walking on a treadmill according to an interval-training pattern and was carried out daily on weekdays for 6 weeks; training intensity and duration were increased weekly. At the end of the training programme, the physical performance of all patients had improved dramatically. The distance walked in 30 min increased from 1.6 to 3.2 km. Furthermore, mean heart rate and lactate concentration at the usual walking speed (5 km/h) decreased significantly from 149 to 120 bpm and from 3.6 to 1.7 mmol/l [8].

In a controlled study, we assessed the effects of an endurance training programme in 36 cancer patients with solid tumours or non-Hodgkin's lymphoma after HDC and autologous peripheral blood stem cell transplantation (PBSCT) [7, 9]. Eighteen patients carried out a daily aerobic exercise programme. Training was started after discharge from the hospital for 6 weeks. A second group of 18 patients did not exercise and was advised to avoid strenuous physical activities. At the beginning and end of the study, the physical performance of all participants was assessed with a treadmill stress test. Furthermore, the fatigue experienced during daily activities was evaluated in a personal interview. In the 6 weeks after discharge, the exercise group had a significantly larger improvement in maximal performance than the control group. Furthermore, the interviews 7 weeks after discharge revealed that 4 patients in the control group (25%) but none in the exercise group experienced fatigue and/or limitations during daily activities due to reduced endurance. These results strongly indicate the need for physical rehabilitation in patients undergoing HDC and PBSCT and suggest that exercise can be useful in preventing fatigue in this group.

In two other studies we have shown that an aerobic exercise programme reduces the loss of physical performance, psychological distress and fatigue in cancer patients undergoing HDC and PBSCT [6, 9]. Furthermore, evidence indicates that physical activity may improve bone marrow regeneration after HDC [7, 9].

In a recent study, Mock et al. [20] evaluated the effects of exercise on fatigue in women receiving radiation therapy for breast cancer. Forty-six women beginning a 6-week radiation therapy participated in a walking exercise programme during treatment. Following random assignment, the subjects in the exercise group maintained an individualized, self-paced, home-based walking exercise programme throughout treatment. The exercise programme consisted of a brisk 20- to 30-min walk. Subjects walked at their own pace 4–5 times per week, while the control group received the usual care. Fatigue increased in both groups during treatment; however, patients in the exercise programme group had substantially lower fatigue scores than controls. Anxiety, depression and sleep disorders were also observed in both groups; however, the symptom intensity was higher in the usual-care group. To our knowledge, this is the first report about the effects of exercise on radiation-related fatigue. These provocative results suggest that aerobic exercise can be a useful supportive therapy for patients undergoing radiotherapy.

## Conclusions and Suggestions for Future Research

The studies discussed show that aerobic exercise can prevent the onset and reduce the intensity of cancer fatigue in patients undergoing chemotherapy. However, little is known about the effects of exercise in cancer patients during and after radiotherapy. So far, there has been no information about the feasibility of structured exercise programmes during radiotherapy and its effects on fatigue, physical performance and psychological distress. Some evidence suggests that fatigue associated with radiotherapy may be due to a systemic reaction to tissue injury and related to an increased release of cytokines as a consequence of tissue necrosis after radiotherapy. However, no studies have evaluated the biochemical correlates of fatigue during radiotherapy.

Finally, information is needed about the feasibility of exercise programmes for different groups of cancer patients. Controlled, randomized studies should evaluate the effect of physical activity in groups of cancer patients at high risk of developing severe or persistent fatigue (i.e. patients with haematological malignancies undergoing myeloablative procedures).

## References


1   Berglund G, Bolund C, Fornander T, Rutqvist LE, Sjoden PO: Late effects of adjuvant chemotherapy and postoperative radiotherapy on quality of life among breast cancer patients. Eur J Cancer 1991;27:1075–1081.

2   Braith RW, Welsch MA, Mills RM, Keller JW, Pollock ML: Resistance exercise prevents glucocorticoid-induced myopathy in heart transplant recipients. Med Sci Sports Exerc 1998;30: 483–489.

3   Bye A, Ose T, Kaasa S: Quality of life during pelvic radiotherapy. Acta Obstet Gynecol Scand 1995;74:147–152.

4   De Graeff A, De Leeuw RJ, Ros WJ, Hordijk GJ, Battermann JJ, Blijham GH, Winnubst JA: A prospective study on quality of life of patients with cancer of the oral cavity or oropharynx treated with surgery with or without radiotherapy. Oral Oncol 1999;35:27–32.

5   De Graeff A, De Leeuw RJ, Ros WJ, Hordijk GJ, Blijham GH, Winnubst JA: A prospective study on quality of life of laryngeal cancer patients treated with radiotherapy. Head Neck 1999;21:291–296.

6   Dimeo F, Stieglitz R-D, Novelli-Fischer U, Fetscher S, Keul J: Effects of aerobic exercise on the fatigue and psychological status of cancer patients during chemotherapy. Cancer 1999;85: 2273–2277.

7   Dimeo F, Tilmann MH, Bertz H, et al: Aerobic exercise in the rehabilitation of cancer patients after high dose chemotherapy and autologous peripheral stem cell transplantation. Cancer 1997; 79:1717–1722.

8   Dimeo F, Bertz H, Finke J, et al: An aerobic exercise program for patients with haematological malignancies after bone marrow transplantation. Bone Marrow Transplant 1996;18: 1157–1160.

9   Dimeo F, Fetscher S, Lange W, Mertelsmann R, Keul J: Effects of aerobic exercise on the physical performance and incidence of treatment-related complications after high-dose chemotherapy. Blood 1997;90:3390–3394.

10   Forester B, Kornfeld DS, Fleiss JL: Psychotherapy during radiotherapy: Effects on emotional and physical distress. Am J Psychiatry 1985;142:22–27.

11   Forester B, Kornfeld DS, Fleiss JL, Thompson S: Group psychotherapy during radiotherapy: Effects on emotional and physical distress. Am J Psychiatry 1993;150:1700–1706.

12   Graydon JE, Bubela N, Irvine D, Vincent L: Fatigue-reducing strategies used by patients receiving treatment for cancer. Cancer Nurs 1995;18:23–28.

13   Greenberg DB, Sawicka J, Eisenthal S, Ross D: Fatigue syndrome due to localized radiation. J Pain Symptom Manage 1992;7:38–45.

14   Horber FF, Hoppeler H, Scheidegger JR, et al: Impact of physical training on the ultrastructure of midthigh muscle in normal subjects and in patients treated with glucocorticoids. J Clin Invest 1987;79:1181–1190.

15   Kanter MM, Hamlin RL, Unverferth DV, Davis HW, Merola AJ: Effect of exercise training on antioxidant enzymes and cardiotoxicity of doxorubicin. J Appl Physiol 1985;59:1298–1303.

16   King KB, Nail LM, Kreamer K, Strohl R, Johnson JE: Patients description of the experience of receiving radiation therapy. Oncol Nurs Forum 1985;12:55–61.

17   MacVicar MG, Winningham ML, Nickel JL: Effects of aerobic interval training on cancer patients' functional capacity. Nurs Res 1989;38:348–351.

18   Miaskowski C, Lee KA: Pain, fatigue, and sleep disturbances in oncology outpatients receiving radiation therapy for bone metastasis: A pilot study. J Pain Symptom Manage 1999;17:320–332.

19   Mock V, Burke MB, Sheehan P, et al: A nursing rehabilitation program for women with breast cancer receiving adjuvant chemotherapy. Oncol Nurs Forum 1994;21:899–907.

20   Mock V, Dow KH, Meares CJ, et al: Effects of exercise on fatigue, physical functioning, and emotional distress during radiation therapy for breast cancer. Oncol Nurs Forum 1997;24: 991–1000.

21   Monga U, Kerrigan AJ, Thornby J, Monga TN: Prospective study of fatigue in localized prostate cancer patients undergoing radiotherapy. Radiat Oncol Investig 1999;7:178–185.

22   Monga U, Jaweed M, Kerrigan AJ, et al: Neuromuscular fatigue in prostate cancer patients undergoing radiation therapy. Arch Phys Med Rehabil 1997;78:961–966.

23   Smets EM, Garssen B, Schuster-Uitterhoeve A, de Haes J: Fatigue in cancer patients. Br J Cancer 1993;68:220–224.

24   Smets EM, Visser MR, Willems-Groot AF, et al: Fatigue and radiotherapy. A. Experience in patients undergoing treatment. Br J Cancer 1998;78:899–906.

25    Smets EM, Visser MR, Willems-Groot AF, et al: Fatigue and radiotherapy. B. Experience in patients 9 months following treatment. Br J Cancer 1998;78:907–912.

26    Winningham ML, MacVicar MG, Bondoc M, Anderson JI, Minton JP: Effect of aerobic exercise on body weight and composition in patients with breast cancer on adjuvant chemotherapy. Oncol Nurs Forum 1989;16:683–689.

27    Winningham ML, Nail LM, Barton Burke M, et al: Fatigue and the cancer experience: The state of the knowledge. Oncol Nurs Forum 1994;21:23–36.

Fernando Dimeo, MD, Institute of Sports Medicine, Freie Universität Berlin,
Clayallee 229, D–14195 Berlin (Germany)
Tel. +49 30 818 12 575, Fax +49 30 818 12 572, E-Mail ferdimeo@zedat.fu-berlin.de

Dörr W, Engenhart-Cabillic R, Zimmermann JS (eds): Normal Tissue Reactions in Radiotherapy and Oncology. Front Radiat Ther Oncol. Basel, Karger, 2002, vol 37, pp 57–68

# Long-Term Side Effects of Radiotherapy in Survivors of Childhood Cancer

*K. Dieckmann, J. Widder, R. Pötter*

Department of Radiotherapy and Radiobiology, University of Vienna, General Hospital, Vienna, Austria

Thirteen to 14 in 100,000 children will develop a malignancy by the time they are 15 years old [23]. Due to dramatic improvements in survival over the past 20 years, about two thirds of these children can expect to be cured. In Hodgkin's disease [33], acute lymphoblastic leukemia [35] and nephroblastoma [15] 5-year survival rates of about 95% are achieved, whereas for disseminated sarcoma or high grade glioma the 5-year survival rate is less than 25% [30]. In the future 1 in every 900 young adults will be a survivor of a childhood cancer [27].

This success is a tribute to multimodal treatment combining chemotherapy, radiotherapy and surgery. The three modalities complement each other in achieving local and systemic tumor control. At the same time they all lead to various long-term side effects Therefore, the primary goal of tumor control must be complemented by minimization of short and particularly long-term sequelae of treatment. Evidently, increasing long-term cure rates of childhood cancer will be accompanied by an increasing number of patients experiencing long-term side effects of multimodal treatment.

This article focuses on radiotherapy-associated side effects in pediatric oncology also taking account of the contribution of chemotherapy and surgery. In modern treatment protocols radiotherapy is often not the exclusive treatment for childhood tumors, which makes it difficult to clearly identify those side effects which may be attributable to this modality alone. The cornerstone of the assessment of long-term side effects of cancer treatment is its impact on the quality of life. In children and adolescents it is often remarkable how even significant impairment can be integrated into a positive conception of their lives.

## Acute Side Effects of Radiotherapy

Acute side effects of radiation treatment, such as nausea, vomiting, headache, fatigue, mucositis, erythema or hair loss, are mostly transient and depend on the treated volume, dose and fractionation and are not significantly different from the same effects in adults. They are mostly manageable with symptomatic treatment. Potentiation of acute effects of skin or mucous membranes by simultaneous administration of chemotherapy (especially actinomycin D, anthracyclines) should be avoided whenever possible.


## Bone and Soft Tissue

In bone and soft tissue there are virtually no acute radiation-induced side effects. Radiation significantly affects growth in the long term, particularly during maximum growth periods between the 1st and the 6th year of life and around puberty. The decrease relative to target length, thickness and shape of bones (vertebrae, pelvis, head) and target volume and shape of soft tissues due to radiation depends on the age at irradiation, the volume and compartment of tissue irradiated, and on the dose given [14, 32]. The younger the child, the more pronounced growth retardation will be in general. Figure 1 gives an estimate of the amplification of long bones between age 2 and 12 [34]. Irradiation of epiphyses in tubular bones results in an impairment of chondrogenesis and consecutive reduction of the length of the bone, with a different contribution of the proximal and distal epiphysis. The most radiosensitive cells are chondroblasts, while osteoblasts are regarded as less sensitive. An incomplete growth arrest of enchondral ossification is observed at doses of 10–20 Gy, permanent arrest at 20–30 Gy. Perichondral ossification (diaphyses) and desmal ossification (head) are less affected [14]. Myoblasts are as radiosensitive as chondro-blasts; irradiation leads to hypo- or atrophy of involved regions. Higher doses (>50–60 Gy) result in damage to the microvasculature, which may lead to trophic effects in bones and soft tissue, resulting in osteoradionecrosis and fibrosis, respectively [21].

Each vertebra has several centers of ossification in the vertebral body and in the vertebral arch. Given high radiosensitivity, inhomogeneous dose distribution within a vertebra therefore leads to scoliosis or kyphoscoliosis. This especially pertains to unilateral irradiation of abdominal tumors [38], such as neuroblastoma or nephroblastoma in young children and with doses greater than 35 Gy (fig. 2). Scoliosis in such cases is aggravated by concomitant asymmetric hypoplasia of soft tissues. Homogeneous irradiation of vertebrae should therefore always be attempted. The development of symmetric vertebral hypoplasia depends on dose and age of the child [38]. In small children doses of 20–25 Gy

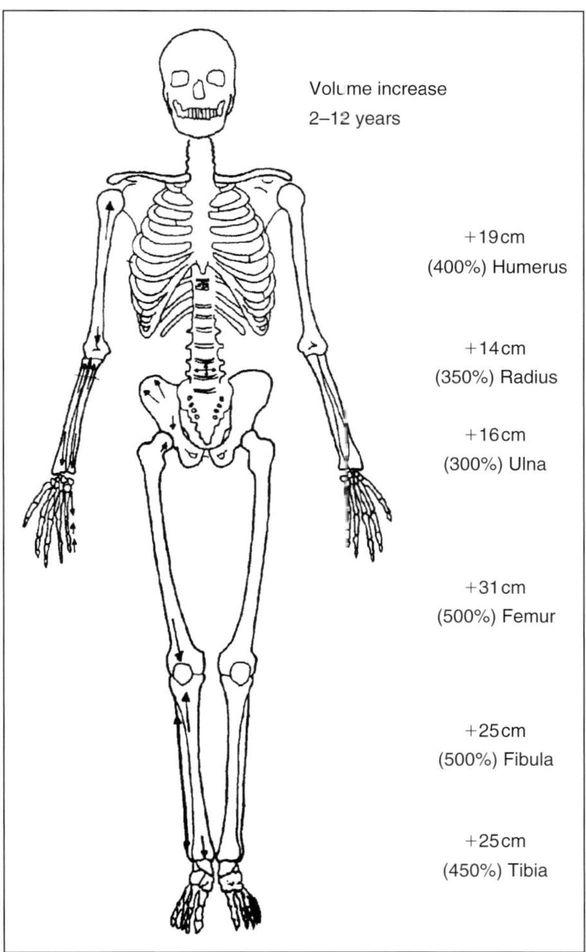


*Fig. 1.* Expected bone growth of various skeletal elements between age 2 and 12 [modified from 34].

lead to a reduced sitting height, whereas the same effect will be observed in older children only at doses ≥35 Gy, especially if large segments of the vertebral column are treated, e.g. for medulloblastoma, acute lymphoblastic leukemia or Hodgkin's disease [32, 39]. After classical mantlefield irradiation for Hodgkin's disease in early childhood with doses >40 Gy, hypoplasia of soft tissues (neck, shoulder girdle, thoracic wall) and bones (vertebral column, mandible, ribs and clavicles) is seen (fig. 3). Radiotherapy doses and treated volumes are therefore steadily being reduced in multimodality treatment for Hodgkin's disease [14, 33].

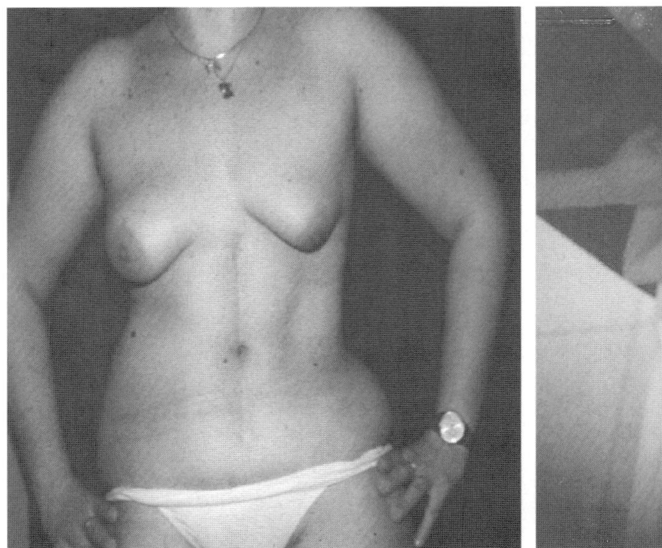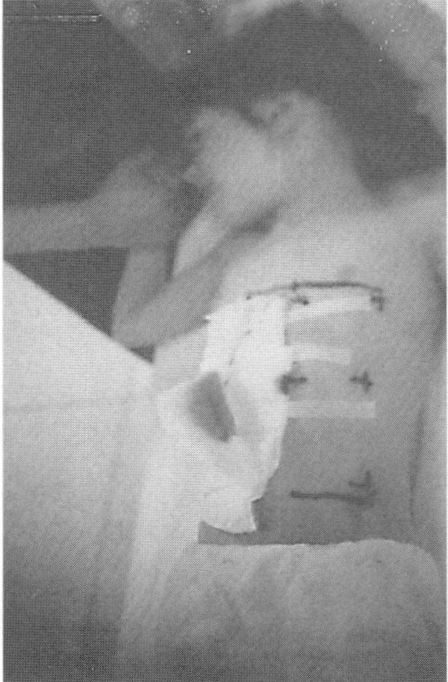

***Fig. 2.*** Nineteen-year-old girl irradiated at the age of 3 years and 10 months because of a Wilms' tumor with a dose of 22.5 Gy. The treatment field is shown on the right. Hypoplasia of the lower part of the left breast, hypothrophy of the left caudal hemithorax and of the left waist are visible on the left.

Early physical therapy often can significantly improve the outlook of children suffering from long-term side effects in connecting tissue (fig. 4). For severe cases, reconstructive surgery may be considered.

### Endocrine System

*Hypothalamic-Pituitary Axis*

The hypothalamic-pituitary region (HPR) is included in radiotherapy fields in whole brain irradiation for leukemia and medulloblastoma, and may be included for other brain tumors, tumors at the scull base, and craniopharyngioma. A wide range of doses is therefore administered. The hypothalamic region appears to be more radiosensitive than the pituitary gland itself, which means that deficiencies of releasing hormones account for most of the hormonal disorders following irradiation of the region. It takes months to years for hormonal disturbances to develop and most of them are irreversible, though amenable to effective treatment if diagnosed early. The most radiosensitive function is growth hormone production. In cranial radiotherapy for leukemia

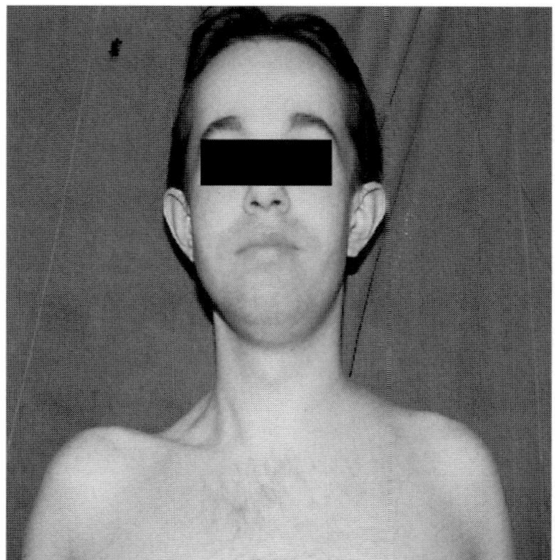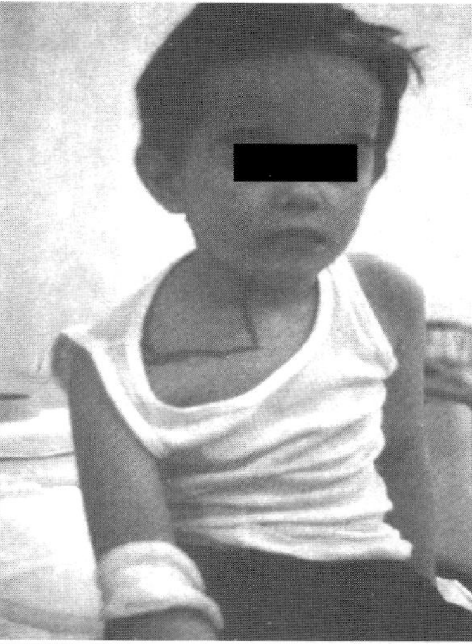

***Fig. 3.*** Twenty-three-year-old boy irradiated at the age of 4 years and 10 months for a non-Hodgkin's lymphoma with a dose of 26 Gy in combination with chemotherapy. On the right photograph the irradiated areas of the right cervical and supraclavicular region are shown. Hypoplasia of the irradiated soft tissue is visible on the left. No functional dysfunction was diagnosed.

with doses of 18–24 Gy (with doses per fraction of 2 Gy), growth retardation is seen, and laboratory analyses reveal abnormal provocation tests in a significant number of patients [6]. A reduction of the dose per fraction to 1.2 Gy with the same total dose has been reported to be followed by normal laboratory findings. If doses ≥45 Gy are administered, all patients will experience growth retardation requiring a substitution of growth hormone [7]. Thyroid-stimulating hormone has been reported to become suppressed in 30–50% of children after 35 Gy to the HPR; in 20–40% of children the follicle-stimulating hormone and luteinizing hormone are also suppressed after the same dose [3, 9]. The production of prolactin, adrenocorticotropin and antidiuretic hormone is more resistant to radiation, rarely giving rise to clinical problems. After any radiotherapy to the HPR with doses ≥18–20 Gy, hormonal screening should be included in the follow-up to detect disorders as soon as possible, to initiate effective substitution.

## Thyroid Gland

Doses of 20–60 Gy result in disturbances of thyroid hormone regulation, which occur with a peak between 3 and 4 years after radiotherapy. Spontaneous

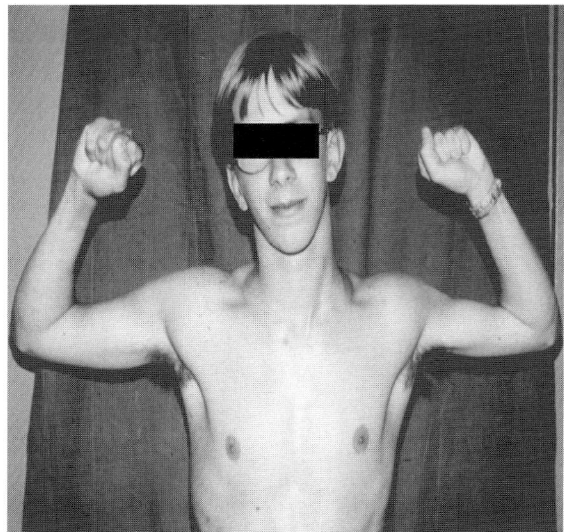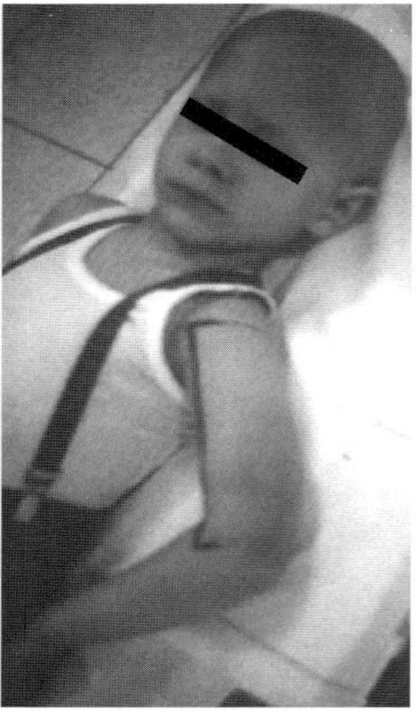

**Fig. 4.** Nineteen-year-old boy irradiated at the age of 3 years and 3 months for a Ewing sarcoma of the left humerus with a dose of 36 Gy plus 10 Gy boost. Only a slight hypotrophy of the soft tissues and bone structures can be noted. No functional restriction can be diagnosed.

normalization of abnormal laboratory findings (elevated TSH) without clinical manifestation is possible [8]. Clinically apparent hypothyroidism is reported in 15–25% of children after 45 Gy to the thyroid; pathological TRH-TSH stimulation tests are seen in 37–78% in this population [8]. TSH levels will be abnormal in 17% of patients after doses of 15–25 Gy [8]. Substitution of thyroid hormones is out of question in clinically manifest hypothyroidism. For chemical hypothyroidism, even if compensated (TSH elevated, thyroid hormones normal), chronic stimulation of the thyroid gland by the elevated TSH may induce further pathological changes in the gland, including thyroid carcinoma. This as yet unproven hypothesis would suggest hormonal substitution even for chemical hypothyroidism.

### Testes

The germinal cells in the testis are much more radiosensitive than the hormone-producing Leydig cells. Total doses <1–2 Gy to the testes, with single doses of <0.1 Gy, lead to a transient disturbance of spermatogenesis, 2–5 Gy are followed by permanent sterility [31]. Alkylating drugs, e.g. high dose cyclophosphamide, also carry a high risk of causing sterility and were replaced by other drugs in some treatment regimens (e.g. Hodgkin's disease in boys, where it was replaced by VP-16 in the German-Austrian study group) [17]. Compromised testosterone production is seen after radiotherapy to the testes for leukemia (24 Gy) [5], but no clear dose-effect relationship has been established until today [4].

*Ovary*

The number of oocytes is determined at birth, postnatally no new oocytes are generated. Lethally damaged oocytes can therefore not be replaced. Beginning in puberty the number of germinal cells is reduced monthly, so that age at radiotherapy is a negative prognostic factor for the maintenance of fertility. If both ovaries receive a dose of 5–10 Gy, permanent sterility will result, even if the treatment took place at a young age [10, 31]. Hormonal disturbance of estrogens and progesterone is seen after somewhat higher doses (10–15 Gy) given to both ovaries. The sparing of one ovary prevents these side effects and should be attempted for any radiotherapy to the pelvis, which sometimes requires surgical oophoropexy. The ovary is less sensitive to alkylating substances than the testis [5]. Hormonal screening with regard to sexual hormones should be performed at regular intervals to prevent disorders of sexual development. Spontaneous improvement of the compromised function is occasionally seen.

## Central Nervous System

Acute side effects of radiotherapy to the whole brain or parts of the brain, such as headache, nausea, vomiting and sleepiness, are usually mild and well manageable with corticosteroids and antiemetic medication. Subacute effects, known as the somnolence syndrome, may arise 3–9 weeks after completion of radiotherapy. Patients present with extreme sleepiness of up to 20 h per day and elevated body temperature; they are anorectic and irritable. This syndrome is interpreted as a transient disturbance of myelinization. It can be mitigated by corticosteroids and is spontaneously reversible. Late side effects are usually irreversible and show a slow progression over months and years. All brain-dependent functions – intellectual, neuropsychological and neurological – may be affected. Cognitive impairment, measured as decreased IQ, has been observed occasionally after 18–24 Gy whole brain irradiation for leukemia, especially when combined with intrathecal methotrexate [19, 29]. Neuropsychological testing, but usually not routine clinical observation, reveals deficits in memory, learning and fine-tuned motor skills at school and in psychosocial adaptation.

Again, in younger children (especially before age 3–5) the impairment will be more severe, higher doses increase the risk (e.g. medulloblastoma with craniospinal doses of 35 Gy). Generalized leukencephalopathy and microangiopathy may arise when higher doses are administered, especially in cases of reirradiation for recurrences. Transversal myelitis with consecutive paralysis, as the most severe long-term side effect in the spinal cord, can usually be avoided by restricting spinal cord doses to 40–45 Gy.

### Heart

Doses of 40 Gy or higher induce pathological proliferation of endothelia with consecutive ischemia and fibrosis of the myocardium, due to a replacement of the damaged intima by myoblasts [18, 36]. This effect is evaluable by echocardiography or scintigraphy and may clinically not be apparent, but cardiac decompensation may occur a long time after treatment. Excluding as much as possible of the heart in mediastinal irradiation (Hodgkin's disease) by inserting a subcarinal shielding block significantly reduces (and possibly avoids) radiation-induced damage to the heart [20]. The best known cardiotoxic chemotherapeutics are anthracyclines, where cumulative doses >300 mg/m² in combination with radiotherapy lead to arrhythmia and cardiomyopathy [37]. High-dose cyclophosphamide – causing acute intramyocardial edema or hemorrhage, with or without pericardial effusion – may enhance anthracycline induced cardiac toxicity in the long term. Early diagnosis (echocardiography) and treatment can prolong the symptom-free period.

### Lung

Doses of 25–40 Gy to small volumes of the lungs rarely result in compromised pulmonary function. Treatment of both lungs with doses of 20 Gy in multiple fractions leads to decreased ventilation and diffusion and a reduction of the vital capacity to 50–75% may occur [1]. Eleven to 14 Gy cause restrictive changes of the lungs [28]. In young children higher more chronic toxicity is observed at lower doses than in older children because of interference with lung and chest wall development in addition to fibrosis and volume loss.

The target cells for radiation-induced injury of the lungs are the pneumocytes type II, fibroblasts and capillary endothelia, leading to desquamation of alveolar epithelia. The predominant late side effect in the lungs is progressive fibrosis of alveolar septa [28].

## Breast

The breast bud is very radiosensitive; 5–10 Gy will cause hypoplasia of the developing breast (fig. 2). It is located around the mamilla and should be shielded whenever possible [16]. Hormonal treatment is not effective for radiation-induced hypoplasia.

## Gastrointestinal System

Acute effects of radiotherapy to the abdomen, particularly when large volumes are treated with high single doses, may be significant and consist of nausea, vomiting, inappetence and acute enteritis with watery, sometimes bloody and mucous stools. Late effects arising from abdominal irradiation are relatively uncommon among survivors of pediatric malignancies. They depend on total dose (>40–50 Gy), volume and site of irradiation. Most often they arise after whole abdominal irradiation. Manifestations of gastrointestinal toxicity include dysphagia, vomiting, abdominal pain, diarrhea, bleeding and anorexia. Intolerance of fat, milk, gluten and fiber-containing food may be observed in these children and cause growth and weight deficits [12]. Pathogenetically, fibrosis develops within the walls of the gastrointestinal tract, with a thickening of serosa, muscularis and submucosa, leading to the malabsorption syndrome, irritable colon or stricture formation. Additionally, fibrosis may be extraintestinal with formation of adhesions, especially in connection with abdominal surgery or chemotherapy. In rare cases an ileus requiring surgery or chronic ulcerations may result as late effects of radiotherapy.

## Liver

Significant late effects in the liver in children are very rare. Tolerance dose of the liver in combination with chemotherapy is about 20 Gy for children; for infants it is assumed to be between 12 and 15 Gy. If veno-occlusive disease or 'radiation hepatitis' occurs shortly after therapy they are most often related to combined treatment of radio- and chemotherapy. Especially after high-dose chemotherapy, veno-occlusive disease can be observed without radiotherapy or with low-dose radiotherapy (e.g. 12 Gy total-body irradiation). Actinomycin D, adriablastin, cytosine in combination with radiotherapy can cause a measurable liver function failure. The regeneration capacity of hepatic tissue is high, so that higher doses (40–45 Gy) can be administered to small volumes of hepatic tissue. Resulting atrophy of these parts is compensated by hypertrophy and hyperplasia, thus preventing chronic dysfunction.

## Secondary Malignancies

Since 1972, the Late Effects Study Group (LESG) has been collecting data of secondary malignancies with pediatric oncology centers throughout the United States, Canada, and western Europe and calculated the actuarial risk of secondary malignancy in childhood cancer survivors. Overall, in children who have experienced a primary malignancy, the incidence of new neoplasms ranges between 3 and 12% at 20 years [11, 26]. The risk is 10–15 times greater than that of age-matched populations. It is highest after cured hereditary retinoblastoma and after Hodgkin's disease, when treated with MOPP chemotherapy and radiotherapy. The most frequent secondary tumors attributable to radiotherapy are sarcomas occurring within the radiation field and breast cancer after radiotherapy for Hodgkin's disease [2, 26]. The development of second malignant neoplasms may be the most serious delayed consequence of successful oncological therapy. The probability of their induction is dose-dependent. Mutations induced by radiation primarily affect dividing cells and lead to clonal expansion, thereby increasing the probability of neoplastic transformation in a multihit pathway. This explains the higher probability of secondary neoplasia in patients who are genetically predisposed, e.g. with hereditary retinoblastoma or osteosarcoma [24, 26], the higher probability for the combination of radiation with certain types of chemotherapy (e.g. alkylating agents), as well as the wide time frame of occurrence from several months to decades after cancer treatment with a mean of 11 years [25].

There are three issues that merit special attention as regards second malignancies after cancer therapy in children. (1) Children surviving cancer have a longer life expectancy than adults or old people, and time is the most important risk factor for the development of a secondary tumor in tissues preexposed to carcinogenic stimuli. (2) The high number of proliferating cells in growing organs allows for an increased chance of DNA changes compared with non-growing tissues. (3) In pediatric oncology higher doses of chemotherapy are used compared with adult cancer treatment due to children's better acute tolerance of cytotoxic drugs. For radiation treatment these three issues indicate that the actual carcinogenic potential of a given dose may be higher for children than for adults or old people.

## References


1   Benoist MR, Lemerle J, Jean R, et al: Effects on pulmonary function of whole lung irradiation for Wilms' tumour in children. Thorax 1982;37:175–180.
2   Bhatia S, Robison LL, Oberlin O, et al. Breast cancer and other second neoplasms after childhood Hodgkin's disease. N Engl J Med 1996;334:745–751.

3  Bieri S, Sklar C, Constine L, Bernier J: Late effects of radiotherapy on the neuroendocrine system. Cancer Radiother 1997;1:706–716.

4  Bramswig JH, Schellong G, Nieschlag E: Pituitary-gonadal function following therapy of testicular relapse in boys with acute lymphoblastic leukemia. Klin Padiatr 1983;195:176–180.

5  Bramswig JH, Korinthenberg R, Gutjahr P: Late effects of antineoplastic chemo- and radiotherapy on the gonads, the nervous system and the occurrence of second malignancies; in Riehm H (ed): Malignant Neoplasms in Childhood and Adolescence. Monographs in Paediatrics. Basel, Karger, 1986, vol 18, pp 84–88.

6  Brennan BM, Rahim A, Mackie EM, Eden OB, Shalet SM: Growth hormone status in adults treated for acute lymphoblastic leukaemia in childhood. Clin Endocrinol (Oxf) 1998;48:777–783.

7  Clayton PE, Shalet SM: Dose dependency of time of onset of radiation-induced growth hormone deficiency. J Pediatr 1991;118:226–228.

8  Constine LS, Donaldson SS, Mc Dougall IR, Cox RS, Link MP, Kaplan HS: Thyroid dysfunction after radiotherapy in children with Hodgkin's disease. Cancer 1984;53:878–883.

9  Constine LS, Woolf PD, Cann D, et al: Hypothalamic-pituitary dysfunction after radiation for brain tumors. N Engl J Med 1993;328:87–94.

10  Dieckmann K, Plemper C, Brämswig JG, et al: Effects of unilateral (UPI) or bilateral (BPI) pelvic irradiation on ovarian function in girls treated for Hodgkin's disease. Radiother Oncol 1996; 40(suppl 1):42.

11  De Vathaire F, Francois P, Hill C, et al: Role of radiotherapy and chemotherapy in the risk of second malignant neoplasms after cancer in childhood. Br J Cancer 1989;59:792–796.

12  Donaldson SS: Radiation enteritis in children. Cancer 1975;35:1167–1178.

13  Donaldson SS, Kaplan HS: Complications of treatment of Hodgkin's disease in children. Cancer Treat Rep 1982;66:977–982.

14  Eifel PJ, Donaldson AA, Thomas FRM: Response of growing bone to irradiation: A proposed late effects scoring system. Int J Radiat Oncol Biol Phys 1995;31:1301–1307.

15  Flentje M, Weirich A, Graf N, Pötter R, et al: Abdominal irradiation in unilateral nephroblastoma and its impact on local control and survival. Int J Radiat Oncol Biol Phys 1998;40:163–169.

16  Furst CJ, Lundell M, Ahlback SO, Holm LE: Breast hypoplasia following irradiation of the female breast in infancy and early childhood. Acta Oncol 1989;28:519–523.

17  Gerres L, Bramswig JH, Schlegel W, Jurgens H, et al: The effects of etoposide on testicular function in boys treated for Hodgkin's disease. Cancer 1998;83:2217–2222.

18  Gottdiener JS, Katin MJ, Borer JS, Bacharach SL, Green MV: Late cardiac effects of therapeutic mediastinal irradiation: Assessment by echocardiography and radionuclide angiography. N Engl J Med 1983;308:569–572.

19  Griffin TW: White matter necrosis, microangiography and intellectual abilities in survivors of childhood leukemia. Association with central nervous system irradiation and methotrexate therapy; in Gilbert HA, Kagan AR (eds): Radiation Damage to the Nervous System. New York, Raven Press, 1980, pp 155–174.

20  Hancock SL, Donaldson SS, Hoppe RT: Cardiac disease following treatment of Hodgkin's disease in children and adolescents. J Clin Oncol 1993;11:1208–1215.

21  Heyn RM: Late effects of radiotherapy in rhabdomyosarcoma; in Nesbit ME (ed): Late Effects in Successfully Treated Children with Cancer. Clinics in Oncology. London, Saunders, 1985, vol 4, No 2, pp 287–298.

22  Hert J: Das Längenwachstum der Röhrenknochen beim Menschen. Das Aktivitätsverhältnis der Epiphysenknorpel. Anat Anz 1979;106:399–413.

23  Kaatsch P, Haaf G, Michaelis J: Childhood malignancies in Germany – Methods and results of a nationwide registry. Eur J Cancer 1995;31A:993–999.

24  Kony SJ, de Vathaire F, Chompret A, et al: Radiation and genetic factors in the risk of second malignant neoplasms after a first cancer in childhood. Lancet 1997;350:91–95.

25  Lemerle J, Oberlin O, de Vathaire F, Pein F, Aubier F: Late and very late effects of therapy – Towards lifetime follow-up of cured patients; in Cancer in Children. Oxford University Press, 1998, pp 84–98.

26  Meadows A, Baum E, Fossati-Bellani F, et al: Second malignant neoplasms in children: An update from the Late Effect Study Group. J Clin Oncol 1985;3:532–538.

27 Meadows A, Black B, Nesbit ME, et al: American Cancer Society: Workshop on children with cancer. Long term survival. Clinical care, research and education. Cancer 1993;71:3213–3215.

28 McDonald S, Rubin P, Phillips TL, Marks LB: Injury to the lung from cancer therapy: Clinical syndromes, measurable endpoints, and potential scoring systems. Int J Radiat Oncol Biol Phys 1995;31:1187–1203.

29 Mulhern RK, Fairclough D, Ochs JA: Prospective comparison of neuropsychologic performance of children surviving leukemia who received 18 Gy, 24 Gy or no cranial irradiation. J Clin Oncol 1991;9:1348–1356.

30 Neville HL, Andrassy RJ, Lobe TE, Bagwell CE, et al: Preoperative staging, prognostic factors, and outcome for extremity rhabdomyosarcoma: A preliminary report from the Intergroup Rhabdomyosarcoma Study IV (1991–1997). J Pediatr Surg 2000;35:317–321.

31 Ortin TT, Shostak CA, Donaldson SS: Gonadal status and reproductive function following treatment for Hodgkin's disease in childhood: The Standford experience. Int J Radiat Oncol Biol Phys 1990;19:873–880.

32 Probert JC, Parker BR: The effects of radiation therapy on bone growth. Radiology 1975;114: 155–162.

33 Pötter R: Paediatric Hodgkin's disease. Eur J Cancer 1999;35:1466–1474.

34 Rubin P, Van Houtte P, Constine L: Radiation sensitivity and organ tolerances in pediatric oncology: A new hypothesis. Front Radiat Ther Oncol 1982;16:62–82.

35 Reiter A, Schrappe M, Ludwig WD, Tiemann M, et al: Intensive ALL-type therapy without local radiotherapy provides a 90% event-free survival for children with T-cell lymphoblastic lymphoma: A BFM Group report. Blood 2000;95:416–421.

36 Steinherz LJ, Steinherz PG, Tan C: Cardiac failure and dysrhythmias 6–19 years after anthracycline therapy: A series of 15 patients. Med Pediatr Oncol 1995;24:352–361.

37 Stewart JR, Fajardo LF, Gillette SM, Constine LS: Radiation injury to the heart. Int J Radiat Oncol Biol Phys 1995;31:1205–1211.

38 Willich E, Kuttig H, Pfeil G, Scheibel P: Pathological changes (developmental/growth disturbances) of the spine after radiation therapy of nephroblastomas during early childhood. A retrospective long-term follow-up study in 82 children. Strahlenther Onkol 1990;166:815–821.

39 Willman K, Cox R, Donaldson SS: Radiation induced height rate impairment in pediatric Hodgkin's disease. Int J Radiat Oncol Biol Phys 1994;28:85–92.

Dr. Karin Dieckmann, Department of Radiotherapy and Radiobiology, General Hospital,
Vienna University, Währinger Gürtel 18–20, A–1090 Vienna (Austria)
Tel. +43 1 40400 2665, Fax +43 1 40400 2666, E-Mail karin.dieckmann@str.akh.magwien.gv.at

Dörr W, Engenhart-Cabillic R, Zimmermann JS (eds): Normal Tissue Reactions in Radiotherapy and Oncology. Front Radiat Ther Oncol. Basel, Karger, 2002, vol 37, pp 69–77

# Normal Tissue Reactions during and after Radiochemotherapy

*Jürgen Dunst*

Martin Luther University Halle-Wittenberg, Department of Radiotherapy, Halle, Germany

Complex multimodality treatment concepts play an increasingly more important role in modern oncology. The objectives of these strategies are not only to increase local control and survival but also to preserve the tumor-affected organ and its function and thereby avoid mutilating treatment and assure quality of life. A major disadvantage of complex and combined treatment schedules is the risk of a higher rate of side effects and complications. Knowledge about the nature of side effects with such combined regimes and information on how to deal with them is essential for the oncologist to avoid harm to the patient and to achieve the desired treatment goals.

The objective of this article is to summarize the basic concepts of combined radiochemotherapy regimens with special emphasis on simultaneous radiochemotherapy (sRCT) and to propose general strategies for the prophylaxis and treatment of acute side effects.

## The Concept of Combined Radiochemotherapy: Spatial Cooperation versus Local Enhancement

From a clinical point of view, it is useful to distinguish between two different strategies behind a combination of radiotherapy and chemotherapy. These strategies are best described as 'spatial cooperation' and 'local enhancement' (fig. 1).

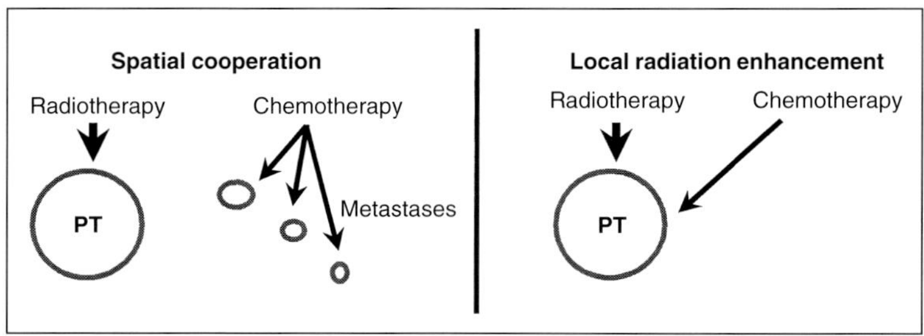


***Fig. 1.*** Concepts of spatial cooperation and local radiation enhancement. PT = Primary tumor.

*Spatial Cooperation*

The concept of spatial cooperation assumes that radiotherapy and chemotherapy work independently of each other at different target sites within the body. Radiotherapy destroys the locoregional tumor and chemotherapy kills occult or clinically detectable metastatic deposits. This concept is very effective if the patient population has a high risk of distant metastases *and* if chemotherapy can effectively treat the metastatic disease. An example for this combination treatment is Ewing's sarcoma. Radiotherapy alone cures about 10% of patients with localized Ewing's sarcoma due to an occult metastatic spread in about 90% of patients. The addition of chemotherapy has during the past 3 decades increased cure rates to about 65% in ongoing studies because combination chemotherapy is highly effective for metastatic disease in Ewing tumors. However, chemotherapy alone would also cure not more than 10% of patients, maybe less, as chemotherapy is almost unable to definitively control the bulky primary tumor. This example, however, clearly demonstrates that under certain conditions the concept of spatial cooperation can be highly effective and can, with regard to cure rates, exert an overadditive effect. This concept of spatial cooperation probably plays the major role in systemic hematological malignancies and in solid tumors with high metastatic potential, e.g. breast cancer, non-small-cell lung carcinoma or pediatric sarcomas.

With regard to sequencing of drugs and radiation as well as to the question of how to manage toxicity it must be emphasized that the dose intensity of chemotherapy often plays a critical role in these concepts. This implies that not only radiotherapy but also chemotherapy must be administered in high doses and in defined intervals. It is therefore often impossible to combine both modalities simultaneously. The optimal sequencing resulting from these considerations is therefore a sequential chemoradiotherapy approach.

*Local Enhancement*

This second concept applies to solid tumors with a predominant locoregional pattern of spread and recurrence, e.g. head and neck, cervix or bladder cancers. Adjuvant or neoadjuvant chemotherapy is nearly ineffective in these cancers. However, the simultaneous combination of chemotherapy and radiotherapy has been shown to significantly increase local control and survival rates. Radiotherapy was in positive studies administered in doses that can be more or less safely combined with a full-dose course of radiotherapy and most studies used single-agent chemotherapy. There was only a minimal decrease in distant metastases. The impact of chemotherapy on survival therefore cannot be explained as a 'systemic' effect rather than an enhancement of the locoregional radiation effect (fig. 1).

To achieve this goal it is necessary to simultaneously combine radiation and chemotherapy. In contrast to the concept of spatial cooperation, chemotherapy for local radiation enhancement must be chosen so that it 'fits' in with radiotherapy. The recent data from several studies in cervical cancer lead to questions of whether the standard approach of medical oncology towards the selection of chemotherapy in a curative treatment setting, namely to use a few drugs and giving them in near-maximum dosage, also applies to sRCT. It is probably better to select only one or two drugs that specifically interact with radiotherapy either by a broad spectrum of efficacy (e.g. cisplatin, taxanes), by specific radiosensitization (e.g. cisplatin, 5-FU, taxanes) or by specifically killing more or less radioresistant cell clones (e.g. mitomycin C).


## Frequency and Course of Toxicity

*Types of Toxicity*

Three different changes of (mainly acute) toxicity result if chemotherapy is added to radiotherapy.

*Additional Nonradiation Toxicity.* Chemotherapy causes specific side effects, especially hematological toxicity and specific organ toxicity depending on the administered drug and dosage. This type of toxicity is uncommon in patients undergoing definitive radiotherapy alone for solid tumors.

*Increased Acute Radiation Toxicity.* In case of sRCT (and to a much lesser degree in case of sequential regimens), there is often a significant increase in acute radiation toxicity, especially mucositis and enteritis.

*New Types of Toxicity.* Due to the simultaneous combination, new and previously unknown or uncommon toxicities may occur. A patient treated with sRCT for head and neck cancers, for example, may develop a long-lasting and severe mucositis which is complicated by long-lasting neutropenia so that the

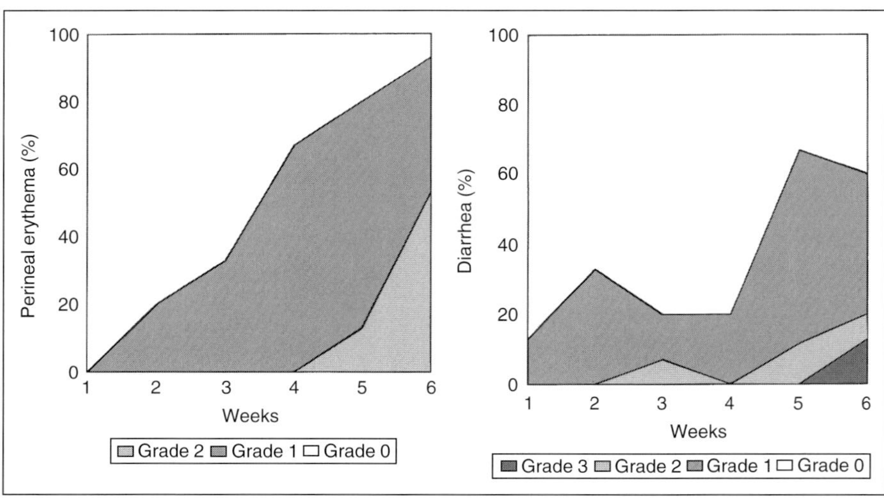


***Fig. 2.*** Perineal erythema and acute diarrhea during postoperative radiochemotherapy in 15 patients with rectal cancer. Patients received 5-FU chemotherapy in weeks 1 and 5 of the treatment schedule (120-hour infusion). There is a peak incidence in the diarrhea score during and shortly after the administration of 5-FU.

risk of local infection and potential subsequent systemic infection, possibly with untypical bacteria, increases. This implies that the prophylaxis and management not only of chemotherapy-related side effects but also of radiation side effects must be performed fast and strictly. It is also questionable whether or not one can adapt intervention recommendations from medical oncology to these situations. It may be sufficient, for example, to start with hematopoietic growth factors in a nonfebrile patient with adjuvant chemotherapy only after the leukocyte count drops below $1,000/mm^3$ but a patient with severe mucositis and ongoing radiotherapy probably requires earlier intervention. However, there are few data and it is difficult to formulate general recommendations.

### Frequency, Degree and Course of Acute Toxicity

The frequency and degree of acute toxicity mainly depends on the sequencing of drugs and radiation. In case of sequential chemotherapy followed by radiotherapy or vice versa, both modalities can mostly be combined without severe problems, especially if radiotherapy follows chemotherapy. This approach should be used in patients with 'systemic' disease. In case of sequential treatment with chemotherapy and radiation, acute radiation toxicity is mostly unchanged. There is, however, some evidence especially from patients with breast cancer that, even when administered separately, the frequency of acute side effects increases as compared to patients treated with radiotherapy alone. A specific question concerns so-called recall phenomena which have been described for several drugs, especially those with radiosensitizing properties such as actinomycin D.

In case of sRCT, acute radiation toxicity may occur more frequently (in more patients) and with an earlier onset [2, 9]. An example for patients treated with adjuvant radiochemotherapy after surgery is given in figure 2 [5]. These patients received additional 5-FU chemotherapy in the 1st and 5th week of a 6-week radiotherapy schedule according to the recommendations of the German Cancer Society. Skin erythema in the radiation field (anal verge) increased steadily over the whole treatment period. In contrast, enteritis as a combination toxicity from both radiation and chemotherapy did also increase but showed a peak incidence in weeks 1 and 2 and 5 and 6 (the weeks with chemotherapy and the weeks thereafter).

*Late Toxicity*

Late effects have, in contrast to acute effects, not been found to be significantly elevated in most regimens with combined radio- and chemotherapy. This may theoretically be explained by the fact that chemotherapy rarely causes severe late damage. In a recent study in head and neck cancer, patients treated with 70 Gy plus chemotherapy had a better local control and survival than patients treated with accelerated radiotherapy alone (77 Gy); in this study, late toxicity was comparable in both arms with a tendency towards less toxicity in patients treated with the combined regimen [Budach, pers. commun.].

However, in situations where late organ damage due to chemotherapy is to be expected radiotherapy may be crucial, and dramatic increases in late toxicity cannot be excluded. An example concerns late cardiac toxicity. Late cardiac damage is mainly a problem of chemotherapy with high doses of anthracyclines or other cardiotoxic drugs. This risk depends on the cumulative dose of drugs. Radiation may also cause late cardiac toxicity and the risk is also dose-dependent. The interaction of both modalities may be described by data from Shapiro et al. [7]. The authors analyzed the risk of cardiac events in patients with adjuvant chemotherapy with or without radiotherapy for breast cancer. Patients received anthracyline-based chemotherapy with two dose levels of Adriamycin and the patients with additional radiotherapy were divided into three subgroups with regard to their cardiac radiation dose (low vs. medium vs. high cardiac dose). High cumulative doses of Adriamycin were associated with a significantly increased risk of cardiac toxicity, whereas radiotherapy alone, irrespective of cardiac dose, was not significant. However, in patients treated with high doses of Adriamycin, the addition of high-dose cardiac radiotherapy dramatically increased the risk of fatal cardiac events (table 1).

*Table 1.* Increased risk of adverse cardiac events in patients mainly depends on anthracycline-based chemotherapy

| Cumulative dose of Adriamycin mg/m² | Radiation dose to the heart | Relative risk for cardiac events | p |
|---|---|---|---|
| 225 | none | 2.1 | n.s. |
| 225 | low | 0 | n.s. |
| 225 | medium | 0 | n.s. |
| 225 | high | 2.3 | n.s. |
| 450 | none | 2.7 | 0.02 |
| 450 | low | 1.7 | 0.9 |
| 450 | medium | 8 | 0.0004 |
| 450 | high | 10 | 0.0001 |

Additional radiotherapy dramatically increases the risk only in patients with high doses of anthracyclines [modified from 7].

## Time Factor

Overall treatment time is a well-known prognostic factor in numerous cancer sites, especially in those that are currently treated with combined radiotherapy and chemotherapy. However, these data result from studies with radiotherapy alone. The time factor is less clear in treatment protocols with combined radio- and chemotherapy. This holds true for patients treated with sRCT for head and neck cancer (table 2) as well as for patients treated with sequential or sRCT for Ewing's sarcoma (table 3).

The interpretation of these data is difficult. However, there is some evidence that a moderate increase in overall treatment time does not decrease the efficacy of radiotherapy if the patient receives additional chemotherapy. This finding has some relevance for the daily clinical practice. If a moderate treatment prolongation is necessary to administer chemotherapy, e.g. if the patients need a treatment break due to acute toxicity before administering the next scheduled chemotherapy course, it is probably better to accept a short treatment delay and go on with the combined radiochemotherapy than to omit chemotherapy.

## Strategies of How to Manage Acute Toxicity

Management of acute toxicity involves not only the treatment, but mainly prophylaxis.

***Table 2.*** Improved locoregional tumor control and survival in patients with sRCT as compared to patients treated with radiation alone despite a significantly higher frequency of treatment breaks and longer treatment time

|  | XRT<br>(n = 140) | XRT + simultaneous<br>chemotherapy<br>(n = 130) |  |
| --- | --- | --- | --- |
| Mucositis III/IV, % | 16 | 38 | $p < 0.001$ |
| Dermatitis III/IV, % | 7 | 17 | $p < 0.05$ |
| XRT breaks | 24/140 | 53/128 | $p < 0.05$ |
| Overall XRT time, days | 51 (10–80) | 53 (40–135) | $p < 0.001$ |
| 3-year locoregional control, % | 17 | 35 | $p < 0.004$ |
| 3-year survival, % | 24 | 49 | $p < 0.0003$ |

XRT = Radiotherapy [from 10].


***Table 3.*** Overall treatment time of radiotherapy (in days) has no impact on local control in patients with Ewing's sarcoma who receive additional chemotherapy during radiotherapy

|  | Scheduled time according to protocol | Patients with local control | Patients with local failure |
| --- | --- | --- | --- |
| Definitive XRT (n = 44) |  |  |  |
|   Duration of chemotherapy prior to start of XRT | 70 | 96 ± 34 | 89 ± 26 |
|   Overall XRT treatment time | 49 | 47 ± 11 | 46 ± 9 |
| Postoperative XRT (n = 82) |  |  |  |
|   Duration of chemotherapy prior to start of XRT | 150 | 164 ± 37 | 169 ± 20 |
|   Overall XRT treatment time | 39 | 37 ± 10 | 37 ± 9 |

XRT = Radiotherapy [modified from 4].


*Adequate Patient Selection*

Patients undergoing combined radiochemotherapy, especially sRCT, should be carefully selected. Patients with a high risk of side effects are not good candidates for aggressive regimens not only because they may develop acute toxicity. Toxicity should be kept below a threshold that makes prolonged breaks necessary or requires termination of radiotherapy after an insufficient dose.

*Toxicity Scoring and Supportive Care*

Patients with sRCT require close observation during treatment because acute toxicity may develop more rapidly than with radiotherapy alone. Clinical examination in short intervals with the scoring of acute toxicity and general condition is mandatory. Patients with weight loss prior to radiochemotherapy or expected long-lasting oral mucositis or esophagitis should receive early enteral feeding via a percutaneous endoscopic gastrostomy [8].

*Early Treatment of Side Effects*

Because acute side effects in radiochemotherapy protocols may develop much more rapidly than with radiation alone, an early start of treatment, if possible, or prophylactic treatment is recommended [10].

*Treatment Breaks*

In case of sRCT, treatment breaks in radiotherapy should be avoided. In case of acute toxicity, chemotherapy may be reduced or delayed during radiotherapy. Short transient breaks in radiotherapy may be justified.

## Impact of Acute Toxicity on Local Control

In some reports in the literature a correlation between acute radiation reaction of normal tissue and improved local tumor control has been established [3, 6]. It is not clear whether this association is a real biological phenomenon. Moreover, the underlying mechanisms are unclear. A genetic background with elevated intrinsic radiosensitivity of normal tissue which is maintained during the development of a tumor has been hypothesized on the basis of animal experiments [1]. Another explanation might be the release of cytokines during an acute radiation-induced inflammation which may change the radiosensitivity of remaining tumor cells in the radiation field. Thus, irrespective of the underlying mechanisms, the occurrence of acute radiation toxicity, as long as it can be managed, is not necessarily an adverse event for long-term outcome.

## References


1   Budach W, Hartford A, Gioioso D, Freeman J, Taghian A, Suit HD: Tumors arising in SCID mice share enhanced radiation sensitivity of SCID normal tissues. Cancer Res 1992;52: 6292–6296.
2   Byhardt RW: Toxicities in RTOG combined-modality trials for inoperable non-small-cell lung cancer. Oncology 1999;13(suppl 5):116–120.
3   Dahl O, Horn A, Mella O: Do acute side-effects during radiotherapy predict tumour response in rectal carcinoma? Radiother Oncol 1994;33:409–413.

4    Dunst J, Jürgens H, Sauer R, Pape H, Paulussen M, Winkelmann W, Rübe C: Radiotherapy for Ewing's sarcoma of bone: An update of the CESS86 data. Int J Radiat Oncol Biol Phys 1995; 32:919–930.

5    Dunst J, Semlin S, Pigorsch S, Müller AC, Reese T: Intermittent use of amifostine during postoperative radiochemotherapy in rectal cancer patients: Impact on acute toxicity and assessment of in vivo radioprotection with the comet assay. Strahlenther Onkol, in press.

6    Kuhnt T, Richter C, Enke H, Dunst J: Acute radiation reaction and local control in breast cancer patients treated with postmastectomy radiotherapy. Strahlenther Onkol 1998;174:257–261.

7    Shapiro CL, Hardenbergh PH, Gelman R, et al: Cardiac effects of adjuvant doxorubicin and radiation therapy in breast cancer patients. J Clin Oncol 1998;16:3493–3501.

8    Thiel HJ, Fietkau R, Sauer R: Malnutrition and the role of nutritional support for radiation therapy patients. Recent Results Cancer Res 1988;108:205–226.

9    Trotti A: Toxicity in head and neck cancer: A review of trends and issues. Int J Radiat Oncol Biol Phys 2000;47:1–12.

10   Wendt T, Grabenbauer GG, Rödel CM, et al: Simultaneous radiochemotherapy versus radiotherapy alone in advanced head and neck cancer: A randomized multicenter study. J Clin Oncol 1998;16:1318–1324.

11   Zimmermann JS, Kimmig B: Pharmacological management of acute radiation morbidity. Strahlenther Onkol 1998;174(suppl 3):62–65.

Prof. Dr. Jürgen Dunst, Department of Radiotherapy,
Martin Luther University Halle-Wittenberg, Dryanderstrasse 4–7, D–06097 Halle (Germany)
Tel. +49 345 557 4310, Fax +49 345 557 4333, E-Mail juergen.dunst@medizin.uni-halle.de

Dörr W, Engenhart-Cabillic R, Zimmermann JS (eds): Normal Tissue Reactions in Radiotherapy
and Oncology. Front Radiat Ther Oncol. Basel, Karger, 2002, vol 37, pp 78–83

# Gemcitabine (Gemzar®) and Radiotherapy – Is It Feasible?

*R. Wilkowski[a], V. Heinemann[a], C. Stoffregen[b]*

[a] Third Medical Department, University Hospital Grosshadern, Munich, and
[b] Lilly Germany GmbH, Bad Homburg, Germany

The wide clinical experience with the use of radiation has resulted in certain principles of treatment. One of these principles is to combine chemotherapy and radiation in order to increase the therapeutic index by synergistic effects. One of the newer drugs which is known to act synergistically with radiation in vitro is gemcitabine. The pyrimidine derivative of deoxycytidine is incorporated into DNA after phosphorylation to triphosphate (dFdCTP) via deoxycytidine kinase. Incorporation of the monophosphate form of this gemcitabine nucleotide instead of the deoxycytidine nucleotide interrupts DNA replication. This leads to a break in the DNA strand and to cell death.

Gemcitabine has demonstrated antitumor activity in a variety of solid tumors [3, 10], and is different from other chemotherapeutic agents because of its relatively low toxicity (typically, mild fatigue and modest myelosuppression).

### Gemcitabine before Irradiation

Approximately 1,200 (12%) patients treated with gemcitabine in the Lilly Clinical Trial Database (CTD) received radiotherapy after completing the trial. There was no evidence of enhanced radiation toxicity among these patients.

In a CALGB trial [14], patients with unresectable stage III non-small cell lung carcinoma (NSCLC) and no prior radiation received two cycles of gemcitabine plus cisplatin immediately before gemcitabine plus cisplatin and concomitant irradiation. There was no enhanced or unexpected toxicity with the concomitant radiation therapy.

In an EORTC Lung Cancer Group phase II study (08941) [15], patients with stage IIIA NSCLC received three cycles of gemcitabine plus cisplatin as induction therapy; responders were then randomized to surgery or radiotherapy. The investigators concluded that gemcitabine plus cisplatin as induction therapy for irradiation is an effective and well-tolerated regimen if patients with severe pulmonary disorders are excluded. The treatment-free interval (from the end of chemotherapy to the start of radiation) was approximately 14 days.

It is difficult to draw conclusions from the limited data available. Up to now it has been noteworthy that (1) there has been no indication of enhanced radiation toxicity with this sequence (gemcitabine before irradiation), (2) there has also been no indication that a treatment-free interval is required before the administration of radiation, and (3) data from two large clinical trials (CALGB and EORTC) support the continued exploration of gemcitabine plus cisplatin preceding irradiation without major concerns regarding the safety and well-being of the participating patients, although long-term toxicity should be monitored through existing pharmacovigilance procedures.


## Gemcitabine after Irradiation

Approximately 818 (8%) patients in the Lilly CTD received radiation before entering a single-agent or combination gemcitabine trial. The median treatment-free interval – from the end of irradiation to the start of gemcitabine therapy – was 14.5 months (0–44 years). Gemcitabine-enhanced postradiation toxicities (i.e. severe adverse events adjacent to the irradiated area such as dermatitis, rash, mucositis, pneumonitis and diarrhea) were rare and difficult to interpret. Pharmacovigilance monitoring did not indicate any trends.

To summarize the data for the gemcitabine postirradiation sequence, there are no reports in the Lilly CTD of any specific treatment-free interval after the acute radiation toxicity has been resolved. It is advisable, however, to start gemcitabine 1 week after completing radiation therapy or after acute radiation-related toxicity has been resolved.


## Gemcitabine Concomitant with Irradiation

*Preclinical Data*

In addition to its cytotoxic effects, gemcitabine is also a potent radiosensitizer in rodent [12] and a variety of human tumor cell lines [7, 8]. Moreover, sensitization was induced more rapidly at higher gemcitabine doses, and was evident up to 48 h after gemcitabine exposure.

*Clinical Data*

Most of the experience with gemcitabine plus radiotherapy during the clinical trial is from patients with advanced NSCLC; however, the question of the superiority of concomitant radiotherapy using gemcitabine and other agents versus radiotherapy alone remains to be settled. In patients with NSCLC, there appears to be a possible enhancement of efficacy with radiation therapy, which may be offset by a risk of increased toxicity or reduced systemic effectiveness.

The first clinical study of concomitant gemcitabine and radiation therapy was conducted by Scalliet et al. [13] in patients with NSCLC. In this phase II study, patients received gemcitabine $1,000 \, mg/m^2$ once weekly concomitantly with radiation up to a maximum dose of 60 Gy (daily fractionation of 2 Gy/day for 5 days/week) for 6 weeks. The planned treatment volume included the visible primary tumor with a 2- to 2.5-cm margin, supraclavicular, ipsilateral/ contralateral hilar, and subcarinal lymph nodes, the inferior mediastinum to the diaphragm including lower lobe lesions, and the entire lung if the lesion was extensive. Eight patients were enrolled before the study was discontinued due to unacceptable esophageal and pulmonary toxicity occurring in 7 patients. Both acute (lung, pharynx/esophagus, skin and upper gastrointestinal) and late (heart and lung) toxicities (RTOG score) were observed in all 8 patients and 6 patients, respectively. Three patients suffered from complications due to acute radiation toxicity (pneumonitis or severe esophagitis), and 2 patients experienced other serious side effects of radiation therapy. There were three treatment-related deaths: two due to pulmonary toxicity and one due to hemorrhage from radiation necrosis. An independent review of the data suggested a correlation between the observed toxicities and the volume of the irradiated tissue. The median treatment volume was $4,795 \, cm^3$ in all 8 patients. Tumor responses consisted of one complete response and five partial responses (four of which were unconfirmed). One patient had stable disease and 1 progressed. Price et al. [11], conducted a phase I dose escalation study to determine the optimal dose and administration schedule of gemcitabine combined with high-dose thoracic irradiation in patients with NSCLC. In the first six dose levels gemcitabine at a dose of $300 \, mg/m^2$ was administered over an increasing number of dose administrations (day 1; days 1 and 15; days 1, 15 and 29; days 1, 8, 15 and 29; days 1, 8, 15, 22 and 29; and days 1, 8, 15, 22, 29 and 36, respectively). At level 7 and above, the dose of gemcitabine was increased in increments of $150 \, mg/m^2$. There were 3–6 patients per dose level. The radiation regimen consisted of 2 Gy/fraction administered for 5 days/week, up to a maximum of 60 Gy. Radiotherapy was administered within 2 h following the start of gemcitabine infusion. The planned treatment volume was not permitted to exceed $2,000 \, cm^3$. Acute grade 3 pharyngitis in 1 patient at dose level 3, and grade 3 pneumonitis and esophagitis in 1 patient each at dose level 7 ($450 \, mg/m^2$) established the maximum tolerated dose at 450 mg/m$^2$ gemcitabine plus standard radiotherapy. No grade 3 or 4 late toxicity was observed. The overall response rate was 65% (11/17 patients) with four complete responses and seven partial responses. These data indicate that concomitant gemcitabine and radiotherapy are effective and well tolerated.

In the CALGB 9431 trial, a phase II randomized trial [14], three treatment arms were investigated in patients with unresectable stage III NSCLC and no prior chemo- or radiotherapy: gemcitabine plus cisplatin (n = 63), vinorelbine plus cisplatin (n = 58) and paclitaxel plus cisplatin (n = 60). Patients received two cycles (cycles 1 and 2) of chemotherapy alone as induction therapy and two cycles (cycles 3 and 4) of chemotherapy concomitantly with irradiation (66 Gy) for a total of four cycles. Cisplatin was administered at 80 mg/m$^2$ in all arms and cycles. Gemcitabine was administered at 1,250 mg/m$^2$ on days 1, 8, 22 and 29, and at 600 mg/m$^2$ with radiation on days 43, 50, 64 and 71. Vinorelbine was given at 25 mg/m$^2$ on days 1, 8, 15, 22 and 29, and at 15 mg/m$^2$ with radiation on days 43, 50, 64 and 71. Paclitaxel was administered at 225 mg/m$^2$ on days 1 and 22 and at 135 mg/m$^2$ with radiation on days 43 and 64. Grade 3/4 toxicities for gemcitabine plus cisplatin, vinorelbine plus cisplatin and paclitaxel plus cisplatin during concomitant chemoradiation therapy were neutropenia in 49, 27 and 48% of patients, thrombocytopenia in 55, 0 and 6% of patients, esophagitis in 49, 24 and 35% of patients, and pulmonary reactions (dyspnea) in 11, 15 and 16% of patients, respectively. During induction therapy, the respective incidences for neutropenia were 49, 55 and 50%, and for thrombocytopenia 23, 2 and 0%. The results from the CALGB trial (9431) indicate that the gemcitabine arm was not associated with any different type, including radiation-specific toxicities such as esophagitis and pulmonary effects, or degree of toxicity than that found in the vinorelbine or paclitaxel arms.


## Materials and Methods

We studied the feasibility of innovative therapy combinations in 23 inoperable patients with pancreatic carcinoma and 14 palliative R1-resected patients over the last 3 years. They were treated with one sequential course of systemic chemotherapy (gemcitabine 1,000 mg/m$^2$ on day 1 and cisplatin 50 mg/m$^2$ on day 2 of a biweekly cycle) followed by a simultaneous radiochemotherapy (gemcitabine 300 mg/m$^2$/day 1, 15, 29; 5-FU 350 mg/m$^2$/day; radiation 5 × 1.8 Gy/week to a total dose 45.0 Gy). Surgery was discussed 10 weeks later. During that time patients received two additional courses of chemotherapy. The schedule was feasible. The strategy resulted in 70% partial remissions (16 patients). Ten patients were reevaluated as operable. 5 R0 resections and 3 R1 resections were performed; 2 patients rejected surgery because of their excellent personal results after the therapy. Fourteen patients postoperatively received the therapy described above after R1 resection.

## Results

An interim retrospective analysis reveals a current median survival time of 29.9 months. Historical control data for 5-FU with radiotherapy alone showed a median of 16.5 months [16]. Currently, another approach for the preoperative concept has also confirmed a potential benefit of this concept [2].

## Conclusions

All these data have so far been based on retrospective evaluations. Although the results are promising, the strategy has to be proven in a planned randomized fashion against the current standard treatment with 5-FU and radiochemotherapy.

In conclusion, clinical trials in NSCLC, pancreatic cancer, head and neck cancer and cervical cancer have defined maximum tolerated doses at specific treatment volumes [1, 2, 4–6, 9, 11, 16]. Although interesting data indicate feasibility and efficacy, a limited number of patients have been studied in these trials. Additional trials are needed for a better understanding of the optimal use of gemcitabine and the position in multimodality treatment strategies including radiotherapy.

## References


1    Blackstock AW, Bernard SA, Richards F, et al: Phase I trial of twice-weekly gemcitabine and concurrent radiation in patients with advanced pancreatic cancer. J Clin Oncol 1999;17:2208.
2    Brunner TB, Grabenbauer GG, Kastl S: Phase II-Studie beim Pankreaskarzinom: Simultane Radio-Chemotherapie mit Gemcitabine und Cisplatin. Strahlenther Onkol 1999;175:A153.
3    Comella P, Frasci G, Panza N, et al: Randomized trial comparing cisplatin, gemcitabine, and vinorelbine with either cisplatin and gemcitabine or cisplatin and vinorelbine in advanced non-small-cell lung cancer: Interim analysis of a phase III trial of the Southern Italy Cooperative Oncology Group. J Clin Oncol 2000;18:1451–1457.
4    Eisbruch A, Shewach DS, Urba S, et al: Phase I trial of radiation concurrent with low-dose gemcitabine for head and neck cancer: High mucosal and pharyngeal toxicity. Proc Am Soc Clin Oncol 1997;16:A1377.
5    Eisbruch A, Shewach DS, Bradford CR, et al: Radiation concurrent with gemcitabine for locally advanced head and neck cancer: A phase I study and intracellular drug incorporation study. J Clin Oncol 2001;19:792–799.
6    Hoffman CJ, McGinn CJ, Szarka C, et al: A phase I study of preoperative gemcitabine (GEM) with radiation therapy (RT) followed by postoperative GEM for patients with localized, resectable pancreatic adenocarcinoma (PaC). Proc Am Soc Clin Oncol 1998;17:A1090.
7    Lawrence TS, Eisbruch A, Shewach DS: Gemcitabine-mediated radiosensitization. Semin Oncol 1997; 24: 24–28.
8    McGinn CJ, Shewach DS, Lawrence TS: Radiosensitizing nucleosides. J Natl Cancer Inst 1996; 88:1193–1203.

9   McGinn CJ, Smith DC, Szarka CE, et al: A phase I study of gemcitabine (GEM) in combination
    with radiation therapy (RT) in patients with localized, unresectable pancreatic cancer. Proc Am Soc
    Clin Oncol 1998;17:A1014.
10  Moore M, Andersen J, Burris H, et al: A randomized trial of gemcitabine versus 5FU as first-line
    therapy in advanced pancreatic cancer. Proc Am Soc Clin Oncol 1995;14:199a.
11  Price A, Groen H, Gregor A, et al: Phase I study of weekly gemcitabine as a radiosensitizer in stage
    III non-small cell lung cancer. IASLC, in press.
12  Rockwell S, Grindey GB: Effect of 2',2'-difluorodeoxycytidine on the viability and radiosensi-
    tivity of EMT6 cells in vitro. Oncol Res 1992;4:151–155.
13  Scalliet P, Goor C, Galdermans D, et al: Gemcitabine (gemcitabine) with thoracic radiotherapy –
    A phase II pilot study in chemonaive patients with advanced non-small cell lung cancer (NSCLC).
    Proc Am Soc Clin Oncol 1998;17:A1923.
14  Vokes EE, Leopold KA, Herndon II, et al: A randomized phase II study of gemcitabine or pacli-
    taxel or vinorelbine with cisplatin as induction chemotherapy (Ind CT) and concomitant chemora-
    diotherapy (XRT) for unresectable stage III non-small lung cancer (NSCLC) (CALGB Study
    9431). Proc Am Soc Clin Oncol 1999;18:A1771.
15  Van Zandwijk N, Crino L, Stolbach L, et al: Phase II study of gemcitabine (Gem) plus cisplatin
    (Cis) as induction regimen for patients with stage IIIA non-small cell lung cancer (NSCLC) by the
    EORTC Lung Cancer Cooperative Group (EORTC 08955). Proc Am Soc Clin Oncol 1998;17:
    A1799.
16  Wilkowski R, Heinemann V, Schalhorn A, Dühmke E: Simultane Radio-Chemotherapie des inop-
    erablen Pankreaskarzinoms mit 5-FU und Gemcitabine. Strahlenther Onkol 1999;175:A153.

Dr. med. R. Wilkowski, Department of Radiooncology, University Hospital Grosshadern,
Ludwig-Maximilians-Universität, Marchioninistrasse 15, D–81377 Munich (Germany)
Tel. +49 89 7095 22080, Fax +49 89 7095 8875,
E-Mail ralf.wilkowski@radonc.med.uni-muenchen.de

Dörr W, Engenhart-Cabillic R, Zimmermann JS (eds): Normal Tissue Reactions in Radiotherapy and Oncology. Front Radiat Ther Oncol. Basel, Karger, 2002, vol 37, pp 84–91

# Secondary Malignancies after Multimodality Treatment Regimens

*Reinhard Kodym*

Clinic for Radiotherapy and Radiobiology, Vienna, Austria

Ionizing radiation and many substances used in antineoplastic chemotherapy, like alkylating agents, are carcinogens and therefore have the potential for inducing the same type of disease they have been applied to cure.

The classic view of carcinogenesis distinguishes between tumor induction and tumor promotion. Tumor induction is usually a single, irreversible process which is caused by mutagenic agents, like ionizing radiation or alkylating substances. Tumor induction alone is insufficient to generate malignant neoplasms. To result in cancer it has to be followed by a long period of tumor promotion. During tumor promotion significant cell proliferation is required. Tumor-promoting processes inducing this proliferative stimulation hence include effects of chemicals, like phorbol esters, to permanent mechanical irritation. From the viewpoint of molecular biology, it is during this time of tumor promotion that the premalignant cells gain all the necessary mutations, ranging from external growth factor independence to the production of metalloproteinases required for invasive tumor growth.

From the data of the atomic bomb survivors in Hiroshima and Nagasaki it is known [11] that the incidence of leukemias peaked 5–10 years after radiation exposure, while the number of solid tumors started to increase after 10–15 years, but continued to increase about 30 years after exposure. For patients who have been treated with ionizing radiation or alkylating agents the delay between exposure and clinical occurrence of secondary malignancy seems to be shorter than in the A bomb survivor population. Analysis of the secondary cancer rate in children treated for Ewing's sarcoma showed [6] that leukemias occurred from 1.5 to 8 years after treatment, while solid tumors were found 7–11 years after radiotherapy.

In radiation protection, a linear or at least monotonous relationship between radiation dose and the incidence of secondary tumors is assumed. This model is

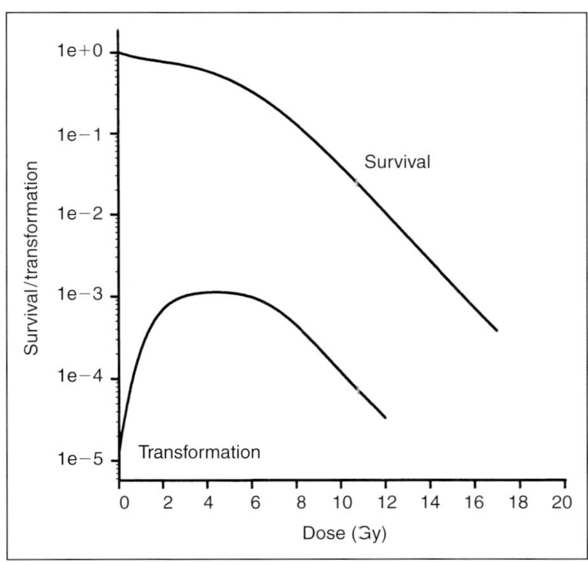



*Fig. 1.* Dose dependence of cell survival and transformation probability of mouse fibroblasts [modified after 8].

not valid for tumor induction by ionizing radiation in the therapeutic dose range. The dose dependence of tumor induction, as obtained form in vitro studies [8], is shown in figure 1. This plot shows the dose-dependent survival of the mouse fibroblast line C3H/10T1/2 with a $D_0$ of about 1.5 Gy. The lower curve represents the probability of cell transformation at various doses. It can be seen that the transformation probability increases with an increasing dose in the low dose range. It reaches its maximum around 4 Gy and drops continuously if the dose is further increased. As ionizing radiation is much more efficient in inducing cell kill than in inducing transformations, potentially transformed cells are inactivated by radiation in the higher dose range. In the in vitro system this results in a continuous decline of transformation frequency once the maximum has been passed. This type of dose response curve is not limited to in vitro experiments. When analyzing the dose effect on tumor induction of electron beams on rat skin, a dose response curve very similar in shape to the in vitro data was reported [1].

## Radiotherapy-Induced Secondary Malignancies

One of the most comprehensive analysis of secondary tumors induced by radiotherapy of a primary tumor was performed by Boice et al. [3, 4], as a

---

case-control study in a very large cohort of women treated for invasive cancer of the uterine cervix. In this cohort, 82,000 patients were treated by external beam irradiation and/or brachytherapy. During the follow-up time of up to 20 years, 3,324 secondary cancers were observed compared to 3,063 cancer cases in the control group. The difference of 261 secondary cancers can be attributed to radiotherapy. Hence, 0.32% of the irradiated patients developed a secondary neoplasm as a consequence of radiotherapy. It is however difficult to distinguish the effect of therapeutic irradiation from the influence of other carcinogens. This is demonstrated by a reduction of the probability for secondary tumors after radiotherapy from 0.32 to 0.15% if all cases of lung tumors in tobacco-smoking patients are excluded from the above-mentioned study.

For this cohort of patients mean organ doses and organ-specific relative risks (RR) for the development of secondary cancer were calculated [4]. Part of these data are shown in figure 2. It can be seen that, as expected, some organs, like bladder or rectum, which received a high dose, also have a significantly higher RR for the development of secondary malignancies. In case of the bladder this RR is 4.0, in case of the rectum it is 1.8. Other organs like colon or uterus seem to be less susceptible to radiation carcinogenesis, with an RR close to 1 despite high doses applied to these organs. In contrast, other organs like stomach and bone marrow received a considerably smaller dose than the organs in the pelvis, and still showed an elevated RR level. Finally, some organs like ovary or connective tissue show a decreased risk for developing malignancies after radiotherapy when compared to a nonirradiated control population. In case of the ovary this is very likely caused by a radiation-induced alteration in the endocrine system.

## Chemotherapy-Induced Secondary Malignancies

The induction of secondary malignancies after cytotoxic chemotherapy depends to a great extent on the specific substances used. Many alkylating agents, like busulfan, cyclophosphamide and phenylalanine mustard, are known to be definitely carcinogenic in humans [7]. Moreover, there is a broad group of substances, belonging to several classes, which are considered to be probably carcinogenic. These include substances with a high cancer induction probability like cisplatin and agents with a low probability like dacarbazine [7]. Finally, there is a third group of substances, mostly antimetabolites like 5-fluorouracil, which are currently considered noncarcinogenic.

The malignancies induced by cytotoxic chemotherapy are usually myelo-dysplastic syndromes (MDS) and acute leukemias (AL). These AL seem to have an extremely poor prognosis when compared to primary leukemias [15].

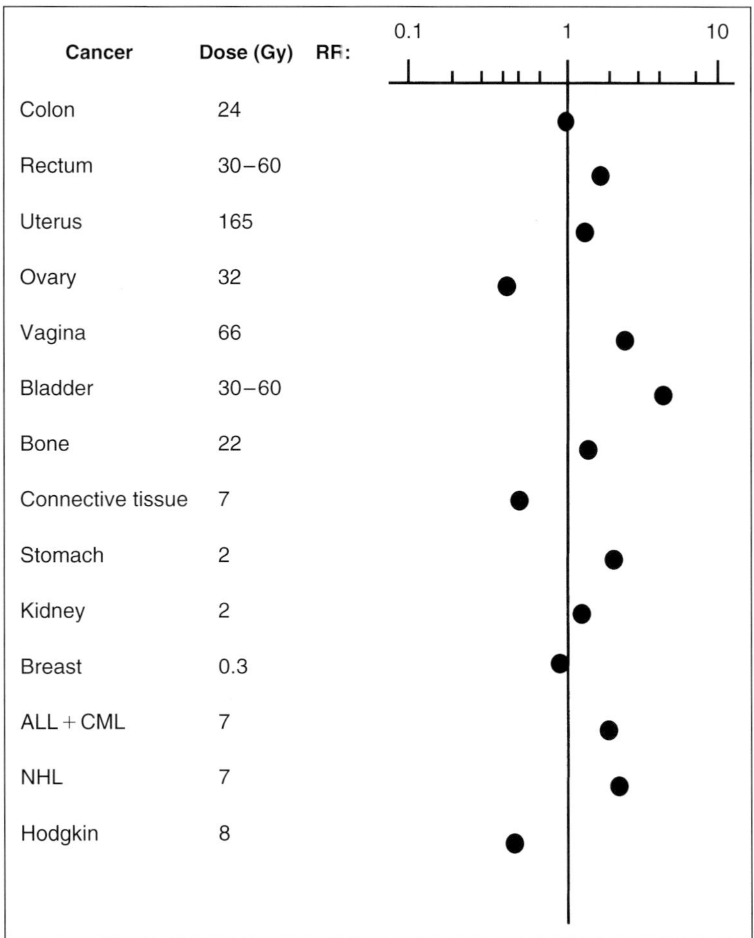


*Fig. 2.* RR for the development of secondary cancers in various organs or organ systems after radiotherapy of invasive cancer of the cervix uteri. Mean dose estimates and RR were taken from Boice et al. [3, 4]. CML = Chronic myelogenous leukemia; NHL = non-Hodgkin's lymphoma.

Analyzing 3,363 patients with ovarian cancer, Green et al. [9] found a cumulative risk of 8.6% for developing MDS or AL within 10 years after successful treatment of ovarian cancer with a regimen containing alkylating substances. RR for the development of MDS or AL after chemotherapy with alkylating agents were published by Curtis et al. [5]. In this study the RR for breast cancer patients of developing MDS or AL was found to be 10.0 when compared to an untreated control group.

It has been shown that the probability of developing cancer increases with an increasing cumulative dose of alkylating agents. In a group of about 400 patients treated for Hodgkin's disease, it was found [14] that the 10-year cumulative risk of developing acute nonlymphatic leukemia was about 6% for patients who receive up to 6 courses of MOPP. The risk doubled for patients receiving 7–12 courses and it rose to somewhat less than 40% if more than 12 cycles of the MOPP regimen were administered.

## Radiochemotherapy-Induced Secondary Malignancies

In the above-mentioned case-control study, Curtis et al. [5] also analyzed the risk for breast cancer patients receiving radiotherapy and radiotherapy in combination with chemotherapy of developing MDS or AL. The RR for patients receiving surgical treatment only was set at 1. Patients receiving radiotherapy only had an RR of 2.4, while patients treated with alkylating agents had a risk of 10.0. For those who received both treatment modalities the risk was found to be 17.4. Hence, the MDS/AL risk for those patients treated with radio- and chemotherapy was considerably higher than the risk associated with each of the treatment modalities alone. It was however lower than one would expect if the carcinogenic potential of radio- and chemotherapy were additive (in this case one would expect a risk of 24). A plausible explanation might be that the pathways on the subcellular level leading to tumor formation by radiotherapy and by chemotherapy are not independent of each other.

There is a marked variation between the risks associated with radiotherapy or chemotherapy for different target organs. Figure 3 shows the RR for several secondary malignancies after treatment of Hodgkin's disease published in [12]. The development of secondary solid tumors in the lung is more likely after radiotherapy (RR 4.3) than after chemotherapy alone (RR 1.1). In contrast, non-Hodgkin's lymphomas are more common after chemotherapy (RR 29.2) than after radiotherapy (RR 9.9). In all cases the RR is greater for the combined modality treatment than for each of the treatment modalities alone, but it usually stays below or at the level of additivity.

The interaction between radiotherapy and chemotherapy in tumor induction is, at least in some cases, more complex than plain overall risk estimation suggests. It has been shown [2] that patients with Hodgkin's disease, who were treated with mantlefield radiotherapy and MOPP chemotherapy, had a cumulative actuarial risk after 14 years for the development of secondary nonlymphatic leukemia of slightly more than 2%. When total nodal irradiation was performed, the cumulative actuarial risk increased to 9%. As all hematopoietic stem cells are likely to be killed in the irradiated volume it can be speculated that the

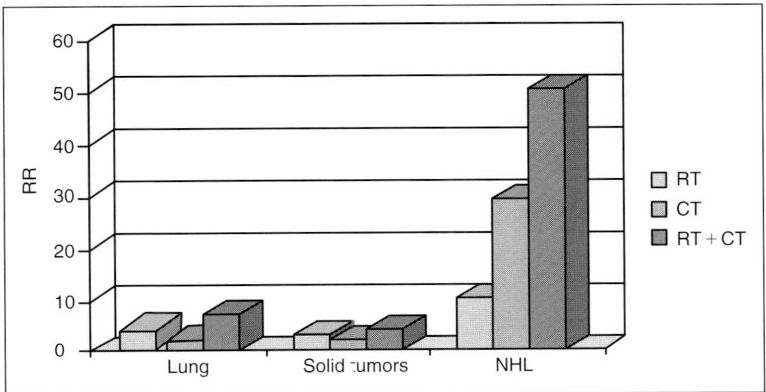



***Fig. 3.*** RR for the development of secondary lung tumors, other solid tumors, and non-Hodgkin's lymphomas (NHL) after treatment of Hodgkin's disease. Left columns indicate the risk after radiotherapy (RT) alone. Middle columns give the risk after chemotherapy (CT) alone, while the right columns show the risk for combined modality treatment (RT + CT) [modified after 12].

increased proliferation of the remaining stem cells after total nodal irradiation results in an increased tumor promotion effect leading to the observed increased risk for nonlymphatic leukemias.

## Secondary Malignancies in Children

The risk of secondary tumor induction by antineoplastic therapy modalities in children deserves special attention for the following reasons.
- Survivors of childhood tumors have the necessary life span of several decades necessary for developing secondary malignancies.
- Many cell systems in children are far from senescence and thus have the proliferative capability necessary for tumor promotion.
- Usually combined modality treatment is used for pediatric tumors.
- There is a much stronger genetic component in many childhood cancers than in malignant tumors of adults.

Hawkins et al. [10] assessed the risk for the induction of secondary bone tumors by combined modality treatment of childhood cancer in a cohort/case-control study involving more than 13,000 patients. Table 1 shows the RR for the development of secondary malignancies of the bone after treatment of various primary cancers. The most probable chromosomal map location of the involved genetic component as well as name of the suspected gene are also given. It can be seen in table 1 that those tumors with a strong genetic component like hereditary

*Table 1.* RR for development of secondary bone cancer after successful treatment of various childhood tumors

| Tumor | RR | Map location (chromosome) | Gene involved |
|---|---|---|---|
| Retinoblastoma (hereditary) | 381 | 13q14.1 | RB1 (LOH) |
| Ewing's sarcoma | 267 | 22q12 t(11;22) | EWS |
| Other bone tumors | 104 | e.g. OS: 13q14.1–2 | ?, Fos |
| Soft-tissue sarcoma | 53 | e.g. RMS: 11p15.5 | ? |
| Hodgkin's disease | 38 | – | |
| Non-Hodgkin's lymphoma | 31 | 18q21.3 t(14;18) | BCL-2 |
| Wilms' tumor | 25 | 11p13 | WT1 |
| Retinoblastoma (non hereditary) | 14 | – | |
| CNS tumors | 12 | – | |
| Leukemia | 5 | – | |

The RR data were published by Hawkins et al. [10]. For each tumor the suspected or proven genetic factor is mentioned. Map locations and gene names were derived from the OMIM database [13].

retinoblastoma or Ewing's sarcoma have a markedly enhanced risk for the development of secondary bone cancer no matter which kind of primary therapy the patients received.

In the same study [10] the RR for secondary bone cancer increased with the dose of ionizing radiation up to a maximum between 30 and 50 Gy (RR 93.4) and then decreased for doses above 50 Gy (RR 64.7). In contrast to this non-monotonous dose-response observed for ionizing radiation, a more or less linear increase of the secondary bone tumor risk was found with increasing doses of alkylating agents.

## Conclusion

The risk for the induction of secondary malignancies after radiotherapy, cytotoxic chemotherapy or combined modality treatment is generally low. Radiotherapy alone is associated with an overall secondary cancer risk well below 0.5% [4] in adults. In children multimodality treatment has been shown to have an overall risk for secondary bone cancer slightly lower than 1% [10]. Despite the low overall probability the individual risk of secondary tumor induction has to be considered in treatment decisions because (1) there are subgroups of patients with a higher risk for secondary malignancies and (2) the better the treatment results become for the primary malignancy the more will long-term results be compromised by secondary cancers.

## References


1    Albert RE, Burns FJ, Heimbach RD: Skin damage and tumour formation from grid and sieve patterns of electron and beta radiation in the rat. Radiat Res 1967;30:525–540.
2    Andrieu JM, Ifrah N, Payen C, et al: Increased risk of secondary acute nonlymphocytic leukemia after extended-field radiation therapy combined with MOPP chemotherapy for Hodgkin's disease. J Clin Oncol 1990;8:1148–1154.
3    Boice JD, Blettner M, Kleinerman RA, et al: Radiation dose and leukemia risk in patients treated for cancer of the cervix. J Natl Cancer Inst 1987;79:1295–1311.
4    Boice JD, Engholm G, Kleinerman RA, et al: Radiation dose and second cancer risk in patients treated for cancer of the cervix. Radiat Res 1988;116:3–55.
5    Curtis RE, Boice JD Jr, Stovall M, et al: Risk of leukemia after chemotherapy and radiation treatment for breast cancer. N Engl J Med 1992;326:1745–1751.
6    Dunst J, Ahrens S, Paulussen M, et al: Second malignancies after treatment for Ewing's sarcoma: A report of the CESS studies. Int J Radiat Oncol Biol Phys 1998;42:379–384.
7    Fraser MC, Tucker MA: Second malignancies following cancer therapy. Semin Oncol Nurs 1989;5:43–55.
8    Goodhead DT: Deductions from cellular studies of inactivation, mutagenesis, and transformation; in Boice JD, Fraumeni JF (eds): Radiation Carcinogenesis: Epidemiology and Biological Significance. New York, Raven Press, 1984, pp 369–385.
9    Green MH, Young RC, Gershenson DM: Melphalan may be a more potent leukemogen than cyclophosphamide. Ann Intern Med 1986;105:360–367.
10   Hawkins MM, Wilson LM, Burton HS, et al: Radiotherapy, alkylating agents, and risk of bone cancer after childhood cancer. J Natl Cancer Inst 1996;88:270–278.
11   Kato H, Schull WJ: Studies of the mortality of A-bomb survivors. Radiat Res 1982;90:395–432.
12   van Leeuwen FE, Klokman WJ, Hagenbeek A, et al: Second cancer risk following Hodgkin's disease: A 20-year follow-up study J Clin Oncol 1994;12:312–325.
13   Online Mendelian Inheritance in Man, OMIM (TM). McKusick-Nathans Institute for Genetic Medicine, Johns Hopkins University (Baltimore, MD) and National Center for Biotechnology Information, National Library of Medicine (Bethesda, MD), 2000. World Wide Web URL: http://www.ncbi.nlm.nih.gov/omim/
14   Pedersen-Bjergaard J, Specht L, Larsen SO, et al: Risk of therapy-related leukaemia and preleukaemia after Hodgkin's disease. Relation to age, cumulative dose of alkylating agents, and time from chemotherapy. Lancet 1987;ii:83–88.
15   Zarrabi MH, Rosner F: Second neoplasms in Hodgkin's disease: Current controversies. Hematol Oncol Clin North Am 1989;3:303–318.



Reinhard Kodym, Clinic for Radiotherapy and Radiobiology,
Währinger Gürtel 18–20, A–1090 Vienna (Austria)
Tel. +43 1 40400 2678, Fax +43 1 40400 2666, E-Mail kodyre1@akh-wien.ac.at

Dörr W, Engenhart-Cabillic R, Zimmermann JS (eds): Normal Tissue Reactions in Radiotherapy and Oncology. Front Radiat Ther Oncol. Basel, Karger, 2002, vol 37, pp 92–100

# Long-Term Effects of Platin and Anthracycline Derivatives and Possible Prevention Strategies

*Jörg Thomas Hartmann*

Department of Hematology/Oncology/Immunology,
Eberhard Karls University, Tübingen, Germany

Cisplatin-based chemotherapy is applied against a variety of solid tumors. Side effects of its application include nephrotoxicity, ototoxicity, neurotoxicity, gastrointestinal sequelae and, to a lower degree, myelosuppression [7, 11, 17, 25]. Cisplatin is usually administrated in combination chemotherapy regimens at a cumulative dose of about $50–100\,mg/m^2$ for 4–6 cycles.

In parallel with the advances in the management of metastatic cancer, growing awareness of late adverse effects of treatment has been shown in numerous publications on long-term survivors, particularly in advanced testicular cancer. Supportive measures have helped to control the acute toxicity of chemotherapy, but the risk of potential long-term alterations remains or even increases with the use of intensive chemotherapy. Systematic evaluation of the late effects of treatment, their extent and reversibility, with special attention to the relationship to the different types of treatment is the premise for the development of regimens with less toxicity. Incorporation of cisplatin into combination regimens not only resulted in high cure rates but also in novel patterns of acute and chronic toxicity, as well as possible prevention strategies of cisplatin-related toxicity. Several publications have demonstrated oto-, neuro- and nephrotoxicity following the application of cisplatin and its substantial impact on fertility [15]. A decrease in quality of life, an increased risk for secondary morbidity and the use of economic resources to treat late toxicities may be avoided with the reduction of therapy-related complications. Recently the aminothiol amifostine has been found to prevent the major, dose-limiting toxicities related to the use of cisplatin [20].

The anthracycline antibiotics, doxorubicin and daunorubicin, cause a cumulative, dose-dependent cardiomyopathy (congestive heart failure, CHF). Dexrazoxane, an EDTA derivative, acting as an intracellular chelating agent, has been shown to decrease the incidence of clinical CHF in patients treated with anthracycline agents [28].


## Nephrotoxicity

Cisplatin-induced renal damage may be associated with a variety of histological changes, including acute focal necrosis of the distal convoluted tubules and collecting ducts, dilatation of convoluted tubules, and the formation of casts. Hyperhydration and forced diuresis have dramatically reduced the incidence of renal complications of cisplatin [26]. However, a persistent 20–30% reduction in GFR was demonstrated in long-term follow-up studies, suggesting that these changes induced by cisplatin are irreversible (table 1). Most investigators have reported that the acute decrease in GFR did not progress further during the months to years following therapy, while tubular function seemed to improve [7]. In some studies the severity of persistent renal impairment was correlated with the dose of cisplatin applied [4, 6, 23]. Although cisplatin is responsible for most of the nephrotoxicity observed during the treatment of testicular cancer, other nephrotoxic agents such as ifosfamide or aminoglycoside antibiotics may contribute to this problem. Patients who are treated with dose-escalated regimens (high-dose chemotherapy, HDCT) in the second-line setting after relapse from cisplatin-based chemotherapy are at a higher risk of renal failure [27]. In most HDCT regimens carboplatin and etoposide have been applied. Some investigators have added either ifosfamide or cyclophosphamide. Dose-escalated ifosfamide may increase the renal damage caused by cisplatin or high-dose carboplatin ($>$1,500 mg/m$^2$) [33].

We compared the nephrotoxicity of different cisplatin-based chemotherapy schedules in terms of changes in GFR, serum magnesium levels and urinary marker excretion. The repeated application of single-day cisplatin (50 mg/m$^2$, day$_{1+22}$) led to a significant decrease in GFR and magnesium levels, while GFR was maintained in patients receiving cisplatin at a daily dose of 20 mg/m$^2$ over 5 consecutive days. However, both groups showed a significant increase in urinary levels of low molecular weight proteins, N-acetyl-β-$D$-glucosaminidase (NAG) and α$_1$-microglobulin, demonstrating that conventional approaches can reduce, but not completely prevent nephrotoxicity. The same investigation using high-dose carboplatin at doses $\geq$1,500 mg/m$^2$ (day$_{1-3}$) has confirmed that its nephrotoxic profile was comparable to 50 mg/m$^2$ cisplatin given as a single-day application [13].

***Table 1.*** Long-term follow-up of renal function after cisplatin-based chemotherapy

| Authors | Patients n | Follow-up months | Median cisplatin dose mg/m² | Mean fall in GFR % |
| --- | --- | --- | --- | --- |
| Fjeldborg et al. [7] | 22 | 16–52 | 452 | 12.5 |
| Daugaard et al. [6] | 30 | n.g. | 1,200[a] | 30 |
| Hansen et al. [10] | 34 | 43–97 | 583 | 8–15 |
| Hamilton et al. [9] | 22 | 8–59 | 400 | 19 |
| MacLeod et al. [22] | 17 | 6–60 | 720[a] | 28 |
| Groth et al. [8] | 25 | 12 | 600 | 29 |
| Bissett et al. [1] | 74 | 13–125 | 400 | 23 |
| Osanto et al. [24] | 36 | 18–112 | 483 | 20 |

GFR represents $^{51}$Cr clearance or creatinine clearance. n.g. = Not given.
[a] Value represents milligrams (absolute).

## Peripheral Neuropathy

The peripheral neuropathy observed in testicular cancer patients is mainly attributed to cisplatin since vinblastine has been replaced by etoposide in standard regimens. The dorsal root ganglion represents the primary target of cisplatin-induced damage. Paresthesia, dysesthesia, disturbances of position, vibratory sensations, and relative sparing of motor units are the clinical signs of neurotoxicity [29]. Up to 76% symptomatic and asymptomatic abnormalities (detected by neurophysiologic testing or vibration threshold) have been reported after chemotherapy for testicular cancer depending on the diagnostic methods [11]. In most investigations acute neurotoxicity disappeared after chemotherapy, but persistent symptoms have been reported in 20–60% of patients. Some studies have identified risk factors for the development of neurotoxicity, such as the cumulative dose of cisplatin given or the simultaneous development of Raynaud's phenomenon. Motor dysfunctions, which occurred rarely during cisplatin therapy, were associated with low serum levels of magnesium.

## Ototoxicity

The reported incidence of ototoxicity, similar to that of neurotoxicity, varies considerably according to the diagnostic methods used. However, investigations using audiometric examinations have revealed an estimated frequency of ototoxicity of approximately 20–40% [2]. Ototoxicity is probably caused by cisplatin damage to the secretory mechanism of the organ of Corti and is manifested as high frequency hearing loss and tinnitus [31]. Regarding risk factors for ototoxicity and its reversibility, the literature is controversial. The cumulative dose of cisplatin applied, its infusional rate, the combination with vinca alkaloids, and preexisting hearing impairment may enhance the risk for the development of ototoxicity [2, 14, 25].

Further aspects of long-term toxicity after cisplatin-based chemotherapy have been alterations of gonadotropin levels (FSH, LH) occurring in up to 60% of patients, and Leydig cell insufficiency persistent in one third of patients. Fertility aspects are also a major concern. Chemotherapy may directly affect the germinal epithelium. Compensated hypogonadism, as described above, is frequent. Recovery appears to peak within 2 years after treatment, occurring more often in younger men. Fertility aspects after treatment may also have an impact on sexual function and quality of life [14].

**Cardiac Toxicity**

The anthracycline doxorubicin causes a cumulative, dose-related cardiomyopathy. Early retrospective studies demonstrated that symptomatic CHF occurred in 6–10% of adults who received cumulative doxorubicin doses $>550 \, mg/m^2$. The frequency of cardiomyopathy can be reduced by modifying the administration schedule [5, 21]. Heart failure was diagnosed in 5% of patients who received a weekly, low-dose schedule of doxorubicin to a cumulative dose $>600 \, mg/m^2$. An increased risk of heart failure was associated with young or old age, 3-weekly application, high cumulative dose, bolus infusion or history of hypertension or coronary heart disease [32]. After cumulative doxorubicin doses of $400 \, mg/m^2$, cardiac monitoring should be frequent. When a dose of $500 \, mg/m^2$ is reached, a monitoring examination should be repeated after every $50 \, mg/m^2$ of doxorubicin.

**Prevention of Toxicity Resulting from Cisplatin-Based Chemotherapy**

Several strategies have been explored to reduce the side effects of treatment, including the use of less intensive treatment or replacement of the nephro- and neurotoxic drug cisplatin by its less toxic analogue carboplatin or changing the schedule of cisplatin administration (e.g. 5 days rather than 1 day). Other approaches to improve the therapeutic index of treatment have included measures to enhance the sensitivity of malignant cells relative to normal tissue or, alternatively, to reduce toxicity to normal tissues, leaving tumor sensitivity

***Table 2.*** Cytoprotectants in cancer treatment [34]

| Agent | Mechanism of action | Clinical use |
| --- | --- | --- |
| Folinic acid | Restores reduced intracellular folate stores | Prevents myelosuppression and mucositis in patients receiving high-dose methotrexate |
| Dexrazoxane | Chelates iron | Protects against anthracycline-induced cardiotoxicity |
| Mesna (mercaptoethane sulfonate) | Active monomer neutralizes cyclophosphamide- and ifosfamide-induced reactive species in urine | Prevents hemorrhagic cystitis |
| Amifostine (WR-2721) | Scavenges free radicals<br>Inactivates reactive species through formation of thioether conjugates<br>Donates hydrogen to DNA radicals | Selectively attenuates toxicity from radiotherapy or chemotherapeutic agents that generate reactive species |

unchanged. Two approaches are under evaluation: first, the administration of the protective agent before chemotherapy or radiation, and second, administration of protective agents following therapy to preferentially rescue normal cells. Table 2 lists currently available chemotherapy and radiation protectors. Amifostine (WR-2721), an organic thiophosphate compound developed in the late 1950s, is a prodrug which is dephosphorylated to its active metabolite, WR-1065, by tissue-bound alkaline phosphate. WR-1065 acts via different mechanisms including radical scavenging, hydrogen donation and, in the case of platinum compounds, prevention or reversal of platinum-DNA adducts. The rationale for the use of amifostine in controlled clinical trials in cancer patients are based on (1) the exclusion of a significant pharmacokinetic interaction between amifostine and cisplatin in the serum [30], (2) the evidence of the reduction of both hematological and nonhematological side effects of cisplatin without evidence of a tumor protection in a large ovarian cancer trial [20] and (3) the identification of cisplatin as a major contributor to acute and long-term toxicity [3].

In a pilot study we have evaluated the degree of kidney damage during cisplatin combination chemotherapy and its possible prevention by amifostine. Patients with different solid tumors were randomized to receive cisplatin ($50\,\text{mg/m}^2$, $\text{day}_1$) as an 1-hour infusion $\pm$ amifostine pretreatment. Amifostine ($910\,\text{mg/m}^2$) was applied as a 15-min infusion prior to the administration of cisplatin. For all patients, creatinine clearance, serum creatinine and electrolytes were determined prior to and after each cycle. Urinary protein and enzyme excretion were measured at days 0, 3 and 5 after chemotherapy. High and low molecular weight proteins, NAG and $\alpha_1$-microglobulin were used to target the glomerular and tubular damage. In 31 evaluable patients, the creatinine clearance remained stable after two cycles of chemotherapy in the amifostine-treated group, while a significant reduction was observed in the control group. The incidence of decreased magnesium levels during treatment was 17% in patients with amifostine compared to 69% in the control arm. The urinary marker excretion was markedly elevated in both groups but indicated more glomerular protection in amifostine-pretreated patients [12].

In addition to preserving renal function in patients treated with highly nephrotoxic regimens (single-day application of $100\,mg/m^2$ cisplatin) or with a sequence of repeated chemotherapy (cumulative cisplatin dose $>400\,mg/m^2$) including high dose chemotherapy [18], further indications for amifostine may be the prevention of other organ toxicities, and particularly pharmacoeconomic aspects. Concerning nephroprotection, questions to be addressed are whether the drug is active in patients who have a reduced renal function before treatment, e.g. after nephrectomy or after intensive pretreatment, and in patients who have a marked reduction of GFR in their early course of treatment.

Furthermore, clinical investigations provided evidence that lower dosages of amifostine ($740\,mg/m^2$) may be sufficient for similar cytoprotection. However, even lower doses with less dose-related side effects and costs may be optimal.

Patients receiving amifostine at a reduced absolute dose of 1,000 mg [corresponding to $540\,mg/m^2$ (500–600) for patients treated in this investigation] prior to cisplatin chemotherapy revealed a slightly lower increase of the tubular urinary markers NAG and albumin. This observation corresponded to a lower incidence of hypomagnesemia in the amifostine group. The use of amifostine at a fixed dosage of 1,000 mg – clinically most relevant – preserved the GFR almost completely in comparison to control patients. It might therefore be appropriate to use the proposed lower dose of amifostine in patients with a restricted number of planned chemotherapy cycles (e.g. $<4$ cycles), a normal GFR at the start of treatment, no preexisting factors for nephrotoxicity, and a body surface area $\leq 2\,m^2$. Additionally, in patients amifostine may result in a low rate of side effects and cost saving [16].

Similar results have been observed in patients with different solid tumors receiving HDCT with carboplatin, etoposide and ifosfamide ($day_{1-3}$: carboplatin $1,500\,mg/m^2$, etoposide $1,500\,mg/m^2$, ifosfamide $12\,g/m^2$) plus peripheral blood stem cell transplantation (PBSC). Forty patients with different solid tumors were randomized to receive HDCT with or without amifostine. Patients who have been treated with amifostine prior to cisplatin revealed a lower degree of renal damage compared to chemotherapy-naive patients. One fourth of patients treated with HDCT alone had a decrease in creatinine clearance of more than 40% from baseline compared to 5% in the amifostine group. The median decrease of serum magnesium levels was 50% higher in patients without amifostine. Compared to the control arm amifostine-treated patients tended to have a lower increase in urinary protein excretion [18].

Reflecting the profile of amifostine protection in chemotherapy-treated patients, these studies indicated, besides the nephroprotective efficacy, a positive effect on mucosal damage and hematological toxicity. The possible protection of long-term neurotoxicity resulting from platin analogues by amifostine is currently being evaluated. The use of amifostine to prevent late toxicities like infertility, vascular and pulmonary toxicity as well as secondary malignancies remains to be addressed in future trials.

Dexrazoxane is an EDTA derivative. The proposed mechanism of cardioprotection is through chelation of intracellular iron, which may decrease doxorubicin-induced free radical generation [28]. The recommended ratio of dexrazoxane to doxorubicin dose is 10:1, and doxorubicin should be given within 30 min of dexrazoxane delivery. It is generally well tolerated. Side effects include pain on injection and myelosuppression.

For the initial use in patients with metastatic breast cancer, three randomized trials involving 625 patients have been conducted. There was a significant reduction in cardiotoxicity. However an increased level of hematological toxicity was seen. A single trial, which was sufficiently powered to detect a significant difference in overall survival, revealed a lower survival rate in dexrazoxane-treated patients. There was no improvement of disease-free or overall survival with the initial use of dexrazoxane in patients with metastatic breast cancer. In contrast, the delayed use in the same group of patients, who have received more than 400 mg/m$^2$ doxorubicin, has revealed significant differences in overall survival and cardiotoxicity in favor of the use of dexrazoxane. However, these data are based on a nonrandomized study including 201 patients. The use of dexrazoxane in an adjuvant pattern, for high-dose anthracycline therapy, other anthracyclines like epirubicin, or in patients with cardiac risk factors is not recommended unless patients are participating in clinical trials which address these questions [19].


## Conclusion

The systematic evaluation of treatment sequelae is an important research field. Cisplatin has been identified to be a major factor of long-term toxicity following curative cancer chemotherapy. After the application of standard dose regimens between 20 and 30% of patients develop cisplatin-related side effects of varying degrees. The indication for the use of amifostine is the prophylactic application prior to cisplatin-based chemotherapy. Monitoring GFR during chemotherapy appears to be sufficient to follow potential renal toxicity on a routine basis. Outside of controlled clinical trials, the use of dexrazoxane is limited to patients with metastatic breast cancer who have received more than 900 mg/m$^2$ doxorubicin and who may benefit from continued anthracycline-containing treatment. Controlled clinical trials are required to further explore whether the use of cytoprotective agents may positively influence the therapeutic ratio in cancer patients.

## References


1 Bissett D, Kunkeler L, Zwanenburg L, Paul J, Gray C, Swan IRC, Kerr DJ, Kaye SB: Long-term sequelae of treatment for testicular germ cell tumours. Br J Cancer 1990;62:655–659.
2 Bokemeyer C, Berger CC, Hartmann JT, Kollmannsberger C, Schmoll HJ, Kuczyk MA, Kanz L: Analysis of risk factors for cisplatin-induced ototoxicity in patients with testicular cancer. Br J Cancer 1998;77:1355–1362.
3 Bokemeyer C, Berger CC, Kuczyk MA, Schmoll HJ: Evaluation of long-term toxicity after chemotherapy for testicular cancer. J Clin Oncol 1996;14:2923–2932.
4 Coleman MP, Esteve J, Damiecki P, Arslan A, Renard H: Trends in cancer incidence and mortality. IARC Sci Publ 1993;121:1–806.
5 Cortes EP, Lutman G, Wanka J, et al: Adriamycin (NSC-123127) cardiotoxicity: A clinicopathologic correlation. Cancer Chemother Rep 1975;6:215–225.
6 Daugaard G, Rossing N, Rorth M: Effects of cisplatin on different measures of glomerular function in the human kidney with special emphasis on high-dose. Cancer Chemother Pharmacol 1988;21:163–167.
7 Fjeldborg P, Sorensen J, Helkjaer PE: The long-term effect of cisplatin on renal function. Cancer 1986;58:2214–2217.
8 Groth S, Nielsen H, Sorensen JB, Christensen AB, Pedersen AG, Rorth M: Acute and long-term nephrotoxicity of *cis*-platinum in man. Cancer Chemother Pharmacol 1986;17:191–196.
9 Hamilton CR, Bliss JM, Horwich A: The late effects of *cis*-platinum on renal function. Eur J Cancer Clin Oncol 1989;25:185–189.
10 Hansen SW, Groth S, Daugaard G, Rossing N, Rorth M: Long-term effects on renal function and blood pressure of treatment with cisplatin, vinblastine, and bleomycin in patients with germ cell cancer. J Clin Oncol 1988;6:1728–1731.
11 Hansen SW, Helweg LS, Trojaborg W: Long-term neurotoxicity in patients treated with cisplatin, vinblastine, and bleomycin for metastatic germ cell cancer. J Clin Oncol 1989;7:1457–1461.
12 Hartmann JT, Fels LM, Knop F, Stolte H, Kanz L, Bokemeyer C: A randomised trial comparing the nephrotoxicity of cisplatin/ifosfamide-based combination chemotherapy with or without amifostine in patients with solid tumors. Invest New Drugs 2000;18:281–289.
13 Hartmann JT, Fels LM, Franzke A, Knop S, Renn M, Maess B, Panagiotou P, Lampe H, Kanz L, Stolte H, Bokemeyer C: Comparative study of the acute nephrotoxicity from standard dose cisplatin ± ifosfamide and high-dose chemotherapy with carboplatin and ifosfamide. Anticancer Res 2000;20:3767–3774.
14 Hartmann JT, Albrecht C, Schmoll HJ, Kuczyk MA, Kollmannsberger C, Bokemeyer C: Long-term effects on sexual function and fertility after treatment of testicular cancer. Br J Cancer 1999; 80:801–807.
15 Hartmann JT, Kanz L, Bokemeyer C: Diagnosis and treatment of patients with testicular germ cell cancer. Drugs 1999;58:257–281.
16 Hartmann JT, Knop S, Fels LM, van Vangerow A, Stolte H, Kanz L, Bokemeyer C: The use of reduced doses of amifostine to ameliorate nephrotoxicity of cisplatin/ifosfamide-based chemotherapy in patients with solid tumors. Anticancer Drugs 2000;11:1–6.

17  Hartmann JT, Kollmannsberger C, Kanz L, Bokemeyer C: Platinum organ toxicity and possible prevention in patients with testicular cancer. Int J Cancer 1999;83:866–869.

18  Hartmann JT, von Wangerow A, Fels LM, Knop S, Stolte H, Kanz L, Bokemeyer C: A randomized trial of amifostine in patients with high dose VIC chemotherapy plus autologous blood stem cell transplantation. Br J Cancer 2001; 84:313–320.

19  Hensley ML, Schuchter LM, Lindley C, Meropol NJ, Cohen GI, Broder G, Gradishar WJ, Green DM, Langdon RJ Jr, Mitchell RB, Negrin R, Szatrowski TP, Thigpen JT, Von Hoff D, Wasserman TH, Winer EP, Pfister DG: American Society of Clinical Oncology clinical practice guidelines for the use of chemotherapy and radiotherapy protectants. J Clin Oncol 1999; 17:3333–3355.

20  Kemp G, Rose P, Lurain J, Berman M, Manetta A, Roullet B, Homesley H, Belpomme D, Glick J: Amifostine pretreatment for protection against cyclophosphamide-induced and cisplatin-induced toxicities: Results of a randomized control trial in patients with advanced ovarian cancer. J Clin Oncol 1996;14:2101–2112.

21  Lefrak EA, Pitha J, Rosenheim S, Gottlieb JA: A clinicopathologic analysis of Adriamycin cardiotoxicity. Cancer 1973;32:302–314.

22  MacLeod PM, Tyrell CJ, Keeling DH: The effect of cisplatin on renal function in patients with testicular tumours. Clin Radiol 1988;39:190–192.

23  Moul JW, Robertson JE, George SL, Paulson DF, Walther PJ: Complications of therapy for testicular cancer. J Urol 1989;142:1491–1496.

24  Osanto S, Bukman A, Van Hoek F, Sterk PJ, De Laat JA, Hermans J: Long-term effects of chemotherapy in patients with testicular cancer. J Clin Oncol 1992;10:574–579.

25  Reddel RR, Kefford RF, Grant JM, Coates AS, Fox RM, Tattersall MH: Ototoxity in patients receiving cisplatin: Importance of dose and method of drug administration. Cancer Treat Rep 1982;66:19–23.

26  Roth BJ, Einhorn LH, Greist A: Long-term complications of cisplatin-based chemotherapy for testis cancer. Semin Oncol 1988;15:345–350.

27  Siegert W, Beyer J, Strohscheer I, Baurmann H, Oettle H, Zingsem J, Zimmermann R, Bokemeyer C, Schmoll HJ, Huhn D: High-dose treatment with carboplatin, etoposide, and ifosfamide followed by autologous stem-cell transplantation in relapsed or refractory germ cell cancer: A phase I/II study. The German Testicular Cancer Cooperative Study Group. J Clin Oncol 1994; 12:1223–1231.

28  Speyer JL, Green MD, Kramer E, Rey M, Sanger J, Ward C, Dubin N, Ferrans V, Stecy P, Zeleniuch-Jacquotte A: Protective effect of the bispiperazinedione ICRF-187 against doxorubicin-induced cardiac toxicity in women with advanced breast cancer. N Engl J Med 1988;319:745–752.

29  Thompson SW, Davis LE, Kornfeld M, Hilgers RD, Standefer JC: Cisplatin neuropathy. Clinical, electrophysiologic, morphologic, and toxicologic studies. Cancer 1984;54:1269–1275.

30  Treskes M, van der Vijgh WJ: WR2721 as a modulator of cisplatin- and carboplatin-induced side effects in comparison with other chemoprotective agents: A molecular approach. Cancer Chemother Pharmacol 1993;33:93–106.

31  van der Hulst RJ, Dreschler WA, Urbanus NA: High frequency audiometry in prospective clinical research of ototoxicity due to platinum derivatives. Ann Otol Rhinol Laryngol 1988;97:133–137.

32  Von Hoff DD, Layard MW, Basa P, Davis HL Jr, von Hoff AL, Rozencweig M, Muggia FM: Risk factors for doxorubicin-induced congestive heart failure. Ann Intern Med 1979;91:710–717.

33  Wagstaff AJ, Ward A, Benfield P, Heel RC: Carboplatin. A preliminary review of its pharmacodynamic and pharmacokinetic properties and therapeutic efficacy in the treatment of cancer. Drugs 1989;37:162–190.

34  Schuchter LM, Merpol N, Glick JH: Radiation and chemotherapy protectors; in DeVita V Jr, Hellman H, Rosenberg SA (eds): Cancer Principles and Practice of Oncology, ed 2, Philadelphia, Lippincott-Raven, 1997, vol 2, chap 65, pp 3087–3092.

J.T. Hartmann, MD, Department of Hematology/Oncology/Immunology, UKT Medical Center II, Eberhard Karls University, Otfried-Mueller-Strasse 10, D–72076 Tübingen (Germany)
Tel. +49 7071 29 82125, Fax +49 7071 29 5689,
E-Mail joerg.hartmann@med.uni-tuebingen.de

Dörr W, Engenhart-Cabillic R, Zimmermann JS (eds): Normal Tissue Reactions in Radiotherapy
and Oncology. Front Radiat Ther Oncol. Basel, Karger, 2002, vol 37, pp 101–111

# Radioprotection of Head and Neck Tissue by Amifostine

*Vratislav Strnad, Rolf Sauer*

Department of Radiation Oncology, University of Erlangen-Nürnberg,
Erlangen, Germany

Patients receiving radiotherapy (RT) for carcinoma of the head and neck
have to accept different side effects of the therapy. Severe oral mucositis occurs
in virtually all patients and involves serious dysphagia over a period of several
weeks and, in the absence of appropriate intensive supporting therapy, severe
malnutrition with all its consequences. At the same time, almost all patients
develop, as a chronic sustained side effect, xerostomia during the first 2 months
after the conclusion of treatment. The underlying salivary gland dysfunction
results in a serious, lifelong restriction of the patient's quality of life.

Amifostine has been shown to possess the highest radioprotective potential
amongst a series of more than 4,000 chemical compounds tested [5]. Also, it
has been associated with an acceptable safety profile [15]. The administration
of this product prior to RT provides significant advantages, since amifostine is
one of the most effective means of protecting normal tissue against radiation-
induced damage [6, 12, 24, 28, 50]. This has been confirmed in numerous trials
[16, 22, 27, 34, 44, 49–51].

In this paper we summarize the results of experimental animal investigations and clinical studies in patients with head and neck cancer undergoing RT
with Amifostine.

## Research Results on Amifostine

### Chemoprotective Effect of Amifostine

Myeloprotection by amifostine has been observed when preceding
the following cytostatic agents: cisplatin, carboplatin, BCNU, melphalan and

cyclophosphamide [3, 4, 11, 27, 47]. Of particular interest for this study are the results of clinical trials employing cisplatin. The results revealed a reduction in adverse events typically associated with cisplatin, such as nephrotoxicity, peripheral neurotoxicity and ototoxicity [4, 11, 27, 49]. No protective effect was observed for tumors.

### Radioprotective Effect of Amifostine

The first clinical trials, aimed at determining the optimum dose of amifostine in RT, were carried out during the 1970s and in the early 1980s [13, 15]. Various doses were tested ranging from 25 to 1,330 mg/m$^2$. A daily dose of 340 mg/m$^2$, given on several days per week with concomitant RT over a 5-week period, showed a very good relative safety and efficacy [15]. The main side effects reported were nausea, vomiting and hypotension.

In a Japanese trial [25], no tumor-protective effects were reported with the administration of amifostine in the case of cervical cancer, nor was there any protective effect for chronic side effects. In several trials with patients with rectal carcinoma, on the other hand, amifostine was found to clearly protect from RT-induced normal tissue damage [16, 19]. These trials clearly demonstrated that the intravenous administration of amifostine 15 min before irradiation resulted in a significantly lower incidence of radiation sequelae, although no protection against tumors was detectable. Studies of the radioprotective effect of amifostine in the salivary glands have demonstrated that this product protects the parotid gland not only against acute, but also against delayed consequences of RT [23, 29, 37, 38]. The dose-modifying factor (DMF) for amifostine preceding RT had a mean of 2.3. Detailed DMF values were as follows: for the weight of the salivary gland 1.9–2.5, for the salivary flow rate 2.9, for the amylase concentration 1.7, for the salivary volume 2.1 and for the duration of secretion 2.1 [23, 37, 38]. These values are high in comparison with those for other organs ($DMF_{intestine} = 1.3$; $DMF_{lung} = 1.2–1.3$; $DMF_{hair} = 1.2–1.7$; $DMF_{skin} = 1.1–1.55$ and $DMF_{testes} = 1.1–1.35$ [33, 39]). The higher DMF values for the parotid gland are most likely related to the salivary glands accumulating the highest concentration of amifostine, as demonstrated in studies on the distribution of amifostine in normal tissues [30, 45].

Salivary gland changes after RT have been investigated by other institutions in animal experiments and in nonrandomized clinical studies. A significant decrease in the secretion rate of the parotid gland has been observed even after exposure to 30 Gy [20] and irreversible changes in the salivary glands were seen after exposure to a total dose of 40 Gy [2]. The reduction of salivary flow is detectable even after just one fraction in the course of treatment [35]. The severity of damage depends on the level of dose administered and the volume of salivary glands irradiated [14, 21, 32, 36]. Despite individual differences, a significant functional restriction is found to occur after RT with a total target dose of 50 Gy. The reduction in the whole salivary flow in this case is in the order of 80–95% [8, 17, 21].

γ-Radiation and corpuscular radiation, even at low doses, may cause lethal damage to serous acinar cells of the salivary gland [41]. The death of these cells is reflected at the clinical level by the acute seroadenosis or sialoadenitis, which is also a chronic consequence of the atrophy of the salivary glands. The early effects of RT result from interphase death (apoptosis) of salivary serous cells while later effects are determined by the ability of surviving cells to repopulate [40, 48]. Following therapeutic doses early degeneration of the serous cells and focal necrosis of the parenchyma occur [7]. The corresponding xerostomia subsequently causes drastic changes in the microflora of the oral cavity [1, 9], which is significantly correlated with an increased risk of oropharyngeal infections [9]. The only current means of preserving salivary gland function is an adequate sparing of parotid gland tissue during RT by restricting the dimensions of the radiation fields or selecting radiotherapeutic techniques which spare at least a proportion of the salivary glands. However, this is rarely possible. Attempts to preserve salivary gland function by medical means up to now have not shown convincing results. In spite of a temporary effect, steroids provide no lasting protection; the administration of neostigmine is also ineffective. The parasympathomimetic agent pilocarpine stimulates the salivary glands outside the radiation field, but has not been found to produce a favorable long-term effect in cases where the salivary glands have been irradiated [10, 18, 31, 46].

Few clinical studies assessing the efficacy of amifostine to protect salivary gland in patients receiving irradiation to the head and neck region are available [22, 26, 42, 43]. Takahashi et al. [26, 42, 43] reported in a randomized trial that amifostine, 100 mg i.v. administered prior to RT, provided radioprotection against mucositis, radiodermatitis and salivary gland injury. Tumor eradication tended to be higher in patients who received amifostine, but this was not significant. The study of McDonald et al. [22] presented changes in flow rates of unstimulated whole saliva, stimulated whole saliva, stimulated parotid saliva and changes in $^{99m}$Tc salivary scintigrams of 9 patients receiving WR-2721. Parotid function significantly recovered to 54% of baseline 18 months after therapy. $^{99m}$Tc salivary scintigrams confirmed this restoration of the parotis function.

A large phase III trial investigating amifostine and RT in head and neck patients has recently been finished [34]. The results are given below.

*Patients and Methods.* From October 1995 to August 1997 315 patients were enrolled in this open multicenter, prospective randomized phase III study comparing amifostine plus standard RT with radiation alone for the treatment of patients with head and neck cancer. Fifteen centers in Europe, 10 centers in the

***Table 1.*** Tumor characteristics

|  | RT + amifostine (n = 150) | RT alone (n = 153) |
|---|---|---|
| *Primary site of disease* |  |  |
| Oropharynx | 77 | 66 |
| Oral cavity | 28 | 33 |
| Larynx | 22 | 24 |
| Hypopharynx | 14 | 15 |
| Nasopharynx | 5 | 6 |
| Unknown | 4 | 9 |
| *T stage* |  |  |
| T0 | 2 | 1 |
| T1 | 25 | 21 |
| T2 | 51 | 53 |
| T3 | 29 | 27 |
| T4 | 38 | 34 |
| TX | 5 | 17 |
| *Nodal status* |  |  |
| N0 | 42 | 46 |
| N1 | 37 | 32 |
| N2 | 68 | 66 |
| N3 | 2 | 8 |
| NX | 1 | 1 |

USA and 4 centers in Canada participated in the study. Twelve patients were randomized but never received treatment and no data other than baseline data were available. The analysis hence includes 303 patients (RT alone 153 patients, RT + amifostine 150 patients). In both arms 80% of patients were males, 20% females; the median age was 55.6 and 56.7 years. The tumor characteristics are summarized in table 1.

Patients with histologically confirmed squamous cell carcinoma of the head and neck region, where at least 75% of each parotid gland was included in the treatment field and who received a total dose of 45 Gy or more to each gland, were included in the study and had undergone definitive or adjuvant RT. Patients were randomized to RT alone or to RT with amifostine 200 mg/m² i.v.

Patients in either treatment arm received daily fractions of 1.8–2.0 Gy for 5 days per week. The median dose of RT was comparable in both treatment arms. Field boundaries, safety borders around the primary tumor and lymph gland metastases were governed by the standard RT principles without major deviations. The primary tumor and the regional lymphatic drainage of the neck were

***Table 2.*** Analysis of acute xerostomia

| RTOG grade | RT + amifostine (n = 149) | RT alone (n = 153) | p value[1] |
|---|---|---|---|
| 0 | 17 (11) | 8 (5) | NS |
| 1 | 57 (38) | 25 (16) | NS |
| 2 | 75 (50) | 120 (78) | <0.0001 |

Figures in parentheses represent percentage.
[1] Pearson $\chi^2$ test.

irradiated during the 1–5 weeks of treatment over two laterally opposing fields up to a total reference dose of $D_{REF}$ = 50 Gy (N0) or $D_{REF}$ = 60 Gy (N+). The primary tumor region was boosted during the last week of treatment with shrinking fields to an envisaged total reference dose of 60–70 Gy. This dose boost was carried out in laterally opposed fields or by means of rotation techniques. The total reference dose ($D_{REF}$) in the region of the primary tumor accordingly amounts to 60–70 Gy over a 6- to 7-week period. The median dose of RT was comparable in both treatment arms: 64 Gy in the RT + amifostine arm and 66 Gy in the RT alone arm. The severity of side effects of RT was determined by the RTOG Acute and Late Radiation Morbidity Scoring Criteria. The severity and duration of mucositis, acute and/or late xerostomia, radiodermatitis, pharyngitis, whole salivary production and, in selected centers, the salivary gland function using $^{99m}$Tc-pertechnate salivary scintigraphy were assessed.

Amifostine in the dose of 200 mg/m$^2$ was administered daily as bolus injection over a 3- to 5-min period daily 15–30 min before each irradiation. The dose of 200 mg/m$^2$ was selected empirically.

*Results.* The primary endpoint for assessing acute xerostomia was the incidence of xerostomia grade 2 or higher, which occurred during or within 1 month following RT. As shown in table 2 acute grade 2 xerostomia occurred in 78% of the patients treated with RT alone. This was reduced by 38% to 50% in the patients treated with RT + amifostine. Table 3 illustrates that amifostine reduced the incidence of grade 2 or higher late xerostomia at 18 months by 50%. The severity of xerostomia was also assessed by the amount of whole saliva (stimulated and unstimulated) that the patient produced. Table 4 displays the proportion of patients with and without significant saliva production (>0.1 g for unstimulated saliva, >0 g for stimulated saliva) at baseline, first follow-up visit and at 1 year. The data show that approximately 1 year after therapy, a significantly greater proportion of patients in the RT + amifostine arm were able to produce a significant amount of saliva without stimulation.

**_Table 3._** Analysis of late xerostomia (18 months)

| RTOG grade | RT + amifostine (n = 67) | RT alone (n = 81) | p value[1] |
|---|---|---|---|
| 0 | 12 (18) | 5 (6) | |
| 1 | 36 (54) | 27 (33) | |
| 2 | 14 (21) | 40 (49) | |
| 3 | 5 (7) | 9 (11) | |
| Grade 2/3 | 19 (28) | 49 (60) | 0.0001 |

Figures in parentheses represent percentage.
[1] Pearson $\chi^2$ test.

**_Table 4._** Whole saliva collected

| | RT + amifostine | RT alone | p value[1] |
|---|---|---|---|
| *Unstimulated saliva* | | | |
| Baseline | | | 0.870 |
| <0.1 g | 16 (11) | 17 (11) | |
| >0.1 g | 132 (89) | 132 (89) | |
| First follow-up visit | | | 0.055 |
| <0.1 g | 33 (28) | 51 (40) | |
| >0.1 g | 84 (72) | 77 (60) | |
| 1 year following radiation | | | 0.004 |
| <0.1 g | 14 (29) | 29 (57) | |
| >0.1 g | 35 (71) | 22 (43) | |
| *Stimulated saliva* | | | |
| Baseline | | | 0.566 |
| <0.1 g | 1 (1) | 2 (1) | |
| >0.1 g | 144 (99) | 144 (99) | |
| First follow-up visit | | | 0.843 |
| <0.1 g | 14 (12) | 16 (13) | |
| >0.1 g | 104 (88) | 110 (87) | |
| 1 year following radiation | | | 0.190 |
| <0.1 g | 10 (20) | 16 (32) | |
| >0.1 g | 39 (80) | 34 (68) | |

Figures in parentheses represent percentage.
[1] Pearson $\chi^2$ test.

**Table 5.** Analysis of acute mucositis

| RTOG grade | RT + amifostine (n = 149) | RT alone (n = 153) | p value[1] |
|---|---|---|---|
| 0 | 9 (6) | 1 (1) | |
| 1 | 24 (16) | 22 (14) | |
| 2 | 64 (43) | 70 (46) | |
| 3 | 47 (32) | 57 (37) | |
| 4 | 5 (3) | 3 (2) | |
| Grade 3/4 | 52 (35) | 60 (39) | 0.438 |

Figures in parentheses represent percentage.
[1] Pearson $\chi^2$ test.

**Table 6.** Analysis of acute mucositis in patients with smaller mucosal volumes in the radiation fields (US and French patients)

| RTOG grade | RT + amifostine (n = 52) | RT alone (n = 57) | p value[1] |
|---|---|---|---|
| 0 | 5 (10) | 1 (2) | |
| 1 | 12 (23) | 3 (5) | |
| 2 | 24 (46) | 31 (54) | |
| 3 | 10 (19) | 22 (39) | |
| 4 | 1 (2) | 0 | |
| Grade 3/4 | 11 (21) | 22 (39) | 0.048 |

Figures in parentheses represent percentage.
[1] Pearson $\chi^2$ test.

With regard to mucositis the primary endpoint was the incidence of grade 3/4 mucositis occurring within 90 days from the start of irradiation. As shown in table 5, there was a trend favoring the amifostine arm with respect to the severity of mucositis, but the differences were not significant. These results were not uniform across treatment centers. Multiple logistic regression analysis show that 'country' and 'total dose of radiation' have significant effects in the model. The analysis of the data from the subset of patients with smaller mucosal volumes in the radiation fields (patients treated in USA and France) showed a statistically significant difference favoring the RT + amifostine group (table 6).

With respect to locoregional control the analyses show no differences between the treatment arms: 63% (52/82) of the patients on the RT + amifostin

*Table 7.* Incidence and severity of treatment-related adverse events associated with amifostine

| Adverse events | RT + amifostine (n = 150) | RT alone (n = 153) |
|---|---|---|
| Nausea | | |
| Grade 3 | 4 (3) | 0 |
| All grades | 64 (43) | 26 (17) |
| Vomiting | | |
| Grade 3 | 8 (5) | 0 |
| All grades | 55 (37) | 10 (6) |
| Hypotension | | |
| Grade 3 | 4 (3) | not measured |
| All grades | 21 (14) | not measured |
| Fever | | |
| Grade 3 | 3 (2) | 0 |
| All grades | 12 (4) | 3 (2) |
| Allergic reaction | | |
| Grade 3 | 3 (2) | 0 |
| All grades | 6 (4) | 1 (1) |
| Fatigue | | |
| Grade 3 | 1 (1) | 0 |
| All grades | 15 (10) | 9 (6) |
| Rigor/chills | | |
| Grade 3 | 0 | 0 |
| All grades | 4 (3) | 1 (1) |
| Hypocalcemia | | |
| Grade 3 | 0 | 0 |
| All grades | 2 (1) | 0 |
| Hiccups | | |
| Grade 3 | 1 (1) | 0 |
| All grades | 2 (1) | 0 |

arm and 64% (56/88) of the patients on the RT alone showed no evidence of local disease at 1 year. Using Kaplan-Meier analyses and log-rank test with a median follow-up of 13 months, 76% of the patient on the RT + amifostine arm and 72% of the patients on the RT alone arm were alive and disease free (p = 0.567) and 89 and 82% were alive (p = 0.263).

Adverse events related to the therapy that occurred more frequently in the RT + amifostine arm are summarized in table 7.

## Conclusion

The actual results of experimental and clinical studies allow the following conclusions.

(1) The administration of amifostine at a dose of $200\,mg/m^2$ as bolus injection prior to each irradiation is safe without significant toxicity.

(2) Amifostine reduces the incidence of xerostomia and improves symptoms in patients undergoing RT for head and neck cancer without reducing the efficacy of RT.

(3) Amifostine probably does not reduce the incidence of mucositis in patients undergoing RT for head and neck cancer.

## References


1    Arcuri MR, Schneider RL: The physiological effects of radiotherapy on oral tissue. J Prosthodont 1992;1:37–41.
2    Bentzen JK: Xerostomia caused by radiotherapy of patients with head and neck cancer. Ugeskr Laeger 1992;154:126–129.
3    Buzaid AC, Murren J, Durivage HJ: High-dose cisplatin plus WR-2721 in a split course in metastatic malignant melanoma. A phase II study. Am J Clin Oncol 1991;13:203–207.
4    Capizzi RL, Scheffler BJ, Schein PS: Amifostine-mediated protection of normal bone marrow from cytotoxic chemotherapy. Cancer 1993;72(suppl):3495–3501.
5    Czerewinski A: Report #MCA1-33 to US Army Medical Research and Development Command, 1972.
6    Davidson DE, Grenan MM, Sweeney TR: Biological characteristics of some improved radioprotectors; in Brady LW (ed): Radiation Sensitizers. Their Use in the Clinical Management of Cancer. New York, Masson, 1980, pp 309–320.
7    De Vicente-Rodriguez JC, Cobo-Plana J, Villa-Vigil MA: Post-radiotherapy xerostomia in patients with oral cancer. Changes in salivary inorganic components and immunoglobulins. Av Odontoestomatol 1991;7:508–509.
8    Dreizen S, Brown LR, Handler S, et al: Radiation induced xerostomia in cancer patients: Effects on salivary and serum electrolytes. Cancer 1976;38:273–278.
9    Epstein JB, Freilich MM, Le ND: Risk factors for oropharyngeal candidiasis in patients who receive radiation therapy for malignant conditions of the head and neck. Oral Surg Oral Med Oral Pathol 1993;76:169–174.
10   Johnson JT, Ferretti GA, Nethery WJ, et al: Oral pilocarpine for post-irradiation xerostomia in patients with head and neck cancer. N Engl J Med 1993;329:390–395.
11   Glover D, Glick J, Weiler C, et al: WR-2721 protects against the hematologic toxicity of cyclophosphamide: A controlled phase II trial. J Clin Oncol 1986;4:584–588.
12   Grdina DJ, Guilford WH, Sigdestad CP, et al: Effects of radioprotectors on DNA damage and repair, proteins and cell-cycle progression. Pharmacol Ther 1988;39:133–137.
13   Grdina DJ, Sigdestad CP: Radiation protectors: The unexpected benefits. Drug Metab Rev 1989; 20:13–42.
14   Kaplan P: Mantle irradiation of the major salivary glands. J Prosthet Dent 1985;54:681–686.
15   Kligerman MM, Glover DJ, Turrisi AT, et al: Toxicity of WR-2721 administered in single and multiple doses. Int J Radiat Oncol Biol Phys 1984;10:1773–1776.
16   Kligerman MM, Turrisi AT, Urtusan RC, et al: Final report on phase I trial of WR-2721 before protracted fractionated radiation therapy. Int J Radiat Oncol Biol Phys 1988;14:1119–1122.

17    Leslie MD, Dische S: The early changes in salivary gland function during and after radiotherapy given for head and neck cancer. Radiother Oncol 1994;30:26–32.

18    LeVeque FG, Montgormery M, Potter D, et al: A multicenter, randomized, double-blind, placebo-controlled, dose-titration study of oral pilocarpine for treatment of radiation-induced xerostomia in head and neck cancer patients. J Clin Oncol 1993;11:1124–1131.

19    Liu T, Liu Y, He S, et al: Use of radiation with or without WR-2721 in advanced rectal cancer. Cancer 1992;69:2820–2825.

20    Marks JE, Vis CC, Gottsman VLL, et al: The effects of radiation on parotid salivary function. Int J Radiat Oncol Biol Phys 1981;7:1013–1019.

21    Marunick MT, Seyedsadr M, Ahmad K, Klein B: The effect of head and neck cancer treatment on whole salivary flow. J Surg Oncol 1991;48:81–86.

22    McDonald S, Meyerowitz C, Smudzin T, Rubin P: Preliminary results of a pilot study using WR-2721 before fractionated irradiation of the head and neck to reduce salivary gland dysfunction. Int J Radiat Oncol Biol Phys 1994;29:747–754.

23    Menard TW, Izutsu KT, Ensign WY, et al: Radioprotection by WR-2721 of gamma-irradiated rat parotid gland: Effect on gland weight and secretion at 8–10 days postirradiation. Int J Radiat Oncol Biol Phys 1984;10:1555–1559.

24    Milas L, Hunter N, Reid BO, Thames HD: Protective effect of S-2(3-amino-propylamino)-ethyl-phosphorothioic acid against radiation damage of normal tissues and a fibrosarcoma in mice. Cancer Res 1982;42:1888–1897.

25    Mitsuhashi N, Takahashi I, Takahashi S, et al: Clinical study of radioprotective effects of amifostine (YM-08310, WR-2721) on long-term outcome for patients with cervical cancer. Int J Radiat Oncol Biol Phys 1993;26:407–411.

26    Niibe H, Takahashi I, Mitsuhashi N: An evaluation of the clinical usefulness of amifostine (YM-08310), radioprotective agent. A double-blind placebo-controlled study. 1. Head and neck tumor. J Jpn Soc Cancer Ther 1985;20:984–993.

27    Peters GJ, van der Wilt CL, Garget F, et al: Protection by WR-2721 of the toxicity induced by the combination of cisplatin and 5-fluorouracil. Int J Radiat Oncol Biol Phys 1992;22:785–789.

28    Phillips TL: Rationale for initial clinical trials and future development of radio-protectors. Cancer Clin Trials 1980;3:165–173.

29    Pratt NE, Sodicoff M, Liss J, et al: Radioprotection of the rat parotid gland by WR-2721: Morphology at 60 days post-irradiation. Int J Radiat Oncol Biol Phys 1980;6:431–435.

30    Rasey JS, Nelson NJ, Mahler PA, Anderson KW, Krohn KA, Menard TW: Radioprotection of normal tissue against gamma rays and cyclotron neutrons with WR-2721: $LD_{50}$ studies and $^{35}$S-WR-2721 biodistribution. Radiat Res 1984;97:598–607.

31    Rieke JW, Hafermann MD, Johnson JT, et al: Oral pilocarpine for radiation-induced xerostomia: Integrated efficacy and safety results from two prospective randomized clinical trials. Int J Radiat Oncol Biol Phys 1995;31:661–669.

32    De Vicente Rodriguez JC, Cobo-Plana J, Villa-Vigil MA: Post-radiotherapy xerostomia in patients with oral cancer. Changes in salivary inorganic components and immunoglobulins. Av Odontoestomatol 1991;7:503–514.

33    Rojas A, Stewart FA, Sorensen JA, et al: Fractionation studies with WR-2721: Normal tissues and tumor. Radiother Oncol 1986;6:51–60.

34    Sauer R, Wannenmacher M, Wasserman T, et al: Randomized phase III trial of radiation ± amifostine in patients with head and neck cancer. Proc ASCO 1999;18:1516.

35    Shannon IL, Trodahl JN, Starcke EN: Radiosensitivity of human parotid gland. Proc Soc Exp Biol Med 1978;157:50–53.

36    Simoric S, Sprem N, Voskresensky I, Racic G: Die Wirkung der Radiotherapie auf die Funktion der Ohrspeicheldrüse. Laryngol Rhinol Otol 1987;66:503–506.

37    Sodicoff M, Conger AD, Pratt NE, Trepper P: Radioprotection by WR-2721 against long-term chronic damage to the rat parotid gland. Radiat Res 1978;76:172–179.

38    Sodicoff M, Conger AD, Trepper P, Pratt NE: Short-term radioprotective effects of WR-2721 on the rat parotid gland. Radiat Res 1978;76:317–326.

39   Stewart FA, Rojas A: Radioprotection of mouse skin by WR-2721 in single and fractionated treatments. Br J Radiol 1982;55:42–47.
40   Stephens LC, Ang KK, Schultheiss TE, et al: Target cell and mode of radiation injury in rhesus salivary glands. Radiother Oncol 1986;7:165–174.
41   Stephens LC, Schultheiss TE, Price RE, et al: Radiation apoptosis of serious acinar cells of salivary and lacrimal glands. Cancer 1991;67:1539–1543.
42   Takahashi I: Clinical use of radioprotector amifostine (YM-08310). J Jpn Med Radiat Soc 1984; 44:1396–1404.
43   Takahashi I, Nagai T, Miyaishi K, et al: Clinical study of the radioprotective effects of Amifostine (YM-08310, WR-2721) on chronic radiation injury. Int J Radiat Oncol Biol Phys 1986;12: 935–938.
44   Tannehill S, Mehta MP: Amifostine and radiation therapy: Past, present and furture. Semin Oncol 1996;23(suppl 8):69–77.
45   Utley JF, Marlowe C, Waddell WJ: Distribution of 35S-labelled WR-2721 in normal and malignant tissues of the mouse. Radiat Res 1976;68:284–291.
46   Valdez IH, Wolff A, Atkinson JC, et al: Use of pilocarpine during head and neck radiation therapy to reduce xerostomia and salivary dysfunction. Cancer 1993;71:1848–1851.
47   Van Laar JA, van der Wilt CL, Treskes M, et al: Effect of WR-2721 on the toxicity and antitumor activity of the combination of carboplatin and 5-fluorouracil. Cancer Chemother Pharmacol 1992; 31:97–102.
48   Vissink A, Down JD, Konings AWT: Contrasting dose rate effects of gamma irradiation on rat salivary gland function. Int J Radiat Biol 1993;61:275–282.
49   Wadler S, Beitler JJ, Rubin JS, et al: Pilot trial of cisplatin, radiation and WR-2721 in carcinoma of the uterine cervix: A New York Gynecologic Oncology Group. J Clin Oncol 1993;11:1511–1516.
50   Yuhas JM, Storer JB: Differential chemoprotection of normal and malignant tissues. J Natl Cancer Inst 1969;42:331–335.
51   Yuhas JM, Spellman JM, Culo FR: The role of WR-2721 in radiotherapy and/or chemotherapy. Cancer Clin Trials 1980;3:211–213.

V. Strnad, MD, Assoc. Prof., Department of Radiation Oncology,
University of Erlangen-Nürnberg, Universitätsstrasse 27, D–91058 Erlangen (Germany)
Tel. +49 9131 8532976, Fax +49 9131 8539335,
E-Mail vratislav.strnad@strahlen.med.uni-erlangen.de

Dörr W, Engenhart-Cabillic R, Zimmermann JS (eds): Normal Tissue Reactions in Radiotherapy and Oncology. Front Radiat Ther Oncol. Basel, Karger, 2002, vol 37, pp 112–120

# Cytohormonal Status and Acute Radiation Vaginitis

*Ingeborg B. Fraunholz*[a]*, Bernhard Schopohl*[a]*, Stefan Falk*[b]*,
Heinz D. Boettcher*[a]

[a] Department of Radiation Oncology, Johann Wolfgang Goethe University and
[b] Pathology Associates, Frankfurt/Main, Germany

The cytohormonal status reflects the maturation of the vaginal epithelium, which proliferates and matures in response to estrogen [3]. It is classified by vaginal smear cytology, a traditional gynecological measurement which is a reliable method for determination of the effects of estrogen on the vaginal epithelium [8]. In the fertile phase or during estrogen therapy a thick epithelium with predominantly eosinophilic superficial cells is found, while in estrogen deficiency, e.g. in senium, there are only one or two layers with predominantly parabasal cells in the smear.

Acute radiation-induced mucosal side effects like acute radiation vaginitis are mainly caused by alterations of the epithelium: based on proliferative impairment of basal cells, an imbalance is created between production and loss of epithelial cells, eventually resulting in partial or complete denudation [17]. The clinical state is often worsened by fungal or bacterial superinfection.

There are no data about interactions between the estrogen effect and the radiation effect on the vaginal epithelium. We hence started this prospective, randomized study to investigate changes of the maturation pattern during radiotherapy and its association with acute radiation vaginitis.

## Materials and Methods

*Patient Characteristics*
Until December 1999 we included 21 patients with endometrial or cervical carcinoma in this prospective study. The medium age was 58 years (range 32–76 years). 13 patients had squamous cell carcinoma of the uterine cervix and 8 patients were diagnosed with

adenocarcinoma of the endometrium. Among the cervix carcinoma patients (medium age 52 years), there were 8 premenopausal and 5 postmenopausal women. All 8 endometrium cancer patients (medium age 68 years) were postmenopausal.

After randomization, 9 patients (6 cancer of the cervix, 3 of the endometrium) received daily estriol vaginal suppositories, starting at least at the beginning of radiotherapy. 4 women were premenopausal and 5 postmenopausal. The (non-estrogen-using) control group included 12 patients, 7 with cervix cancer and 5 with endometrial carcinoma; 4 women were pre-, 8 postmenopausal.

*Radiation Therapy*

All patients with cervix cancer were treated with a primary combined radiotherapy. Teletherapy was delivered after individual three-dimensional treatment planning in a four-field box technique with 6 or 25 MV photon beam by linear accelerator. A total dose of 50.4 Gy in daily fractions of 1.8 Gy, 5×/week, was given to the pelvis. $^{192}$Ir high-dose-rate brachytherapy included a total dose of 42 Gy to point A, delivered in 6 fractions, once a week. After 28 Gy of external therapy an individual shield was used to block the brachytherapy area within the 90% isodose. The total radiation dose at the site of the vaginal epithelium examined was about 50 Gy (from teletherapy).

All patients with endometrial cancer had undergone hysterectomy and were irradiated with vaginal $^{192}$Ir high-dose-rate brachytherapy to the whole length of the vagina with 20 Gy (in 0.5-cm tissue depth) in 3 fractions once a week. The total brachytherapy dose at the vaginal surface was 35–38 Gy. Due to an increased risk of recurrence (infiltration depth in the myometrium) in 3 patients, additional external radiotherapy with the parameters described above was delivered to the pelvis and (lower) para-aortic lymph nodes. Individual shielding was used after 40 Gy. The total dose at the vaginal mucosa was 35–38 Gy from brachytherapy plus about 41 Gy from teletherapy.

*Endpoints and Analysis*

A gynecological examination was performed before the onset of treatment, several (3–6) times during radiotherapy and about 6 weeks after the end of therapy. Each time a vaginal smear was taken from the anterior vaginal wall and was routinely fixed in alcohol ether and stained by the method according to Papanicolaou. Cytological analysis included the assessment of the maturation level according to Schmidt (fig. 1), the type of predominant bacteria and the degree of radiation cell damage. This visual analysis combines several morphological parameters for radiation-induced cellular changes, graded as low, moderate or high.

Acute reactions of the vagina were documented according to modified RTOG/EORTC criteria for oral mucosa (table 1), as there is no particular classification system for acute side effects in the vagina [6].

Statistical analysis was carried out using the Fisher exact test, and probability values less than 0.05 were considered significant.

## Results

Acute radiation vaginitis was diagnosed in 8 patients (about 40%) during 17 gynecological examinations. 5 patients had a grade 1 mucositis, 2 patients a

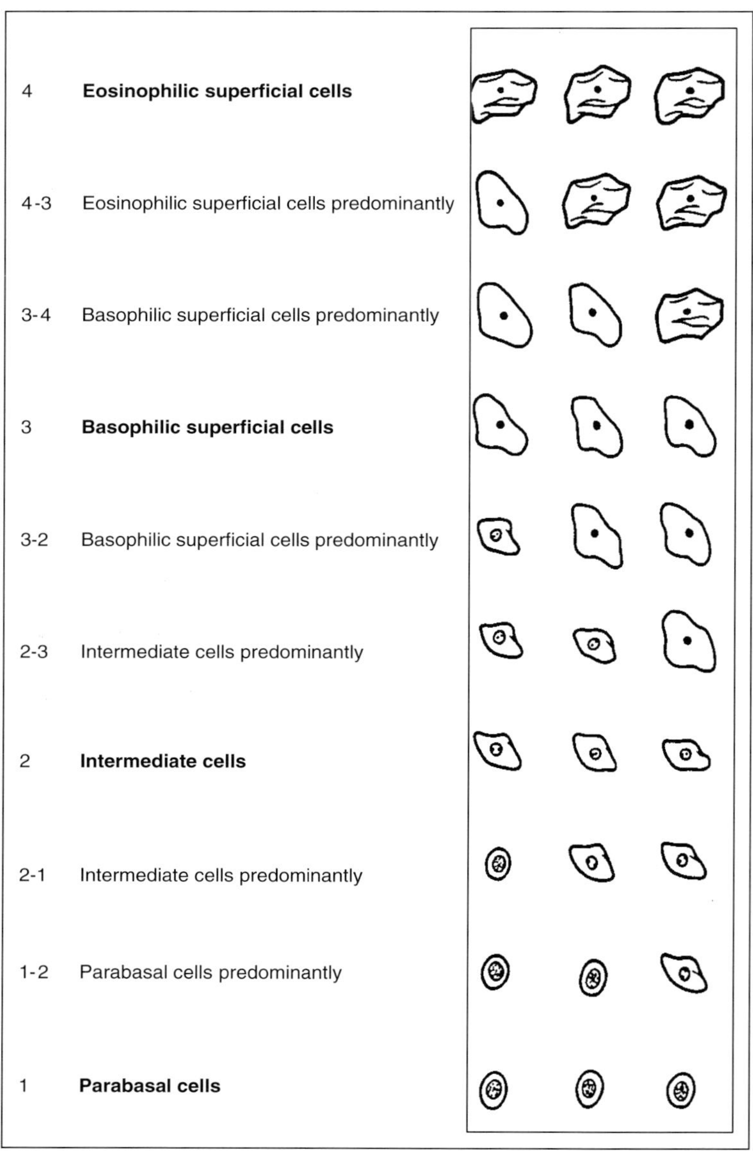



*Fig. 1.* Cytohormonal classification (according to Schmidt).

grade 2 and 1 patient a grade 3 reaction. 2 women (of 8) were premenopausal and 6 (of 13) were postmenopausal. In all 3 patients who were treated with combined radiotherapy for endometrial carcinoma, acute mucosal reactions were seen. In contrast, about 40% acute reactions were found in the group

**_Table 1._** Mucositis-scoring (RTOG/EORTC)

| Grade | Description |
| --- | --- |
| 0 | no acute reaction |
| 1 | erythema; mild pain; no analgesia |
| 2 | patchy mucositis; moderate pain; analgesia |
| 3 | confluent mucositis; severe pain; narcotics |
| 4 | ulceration; hemorrhage; necrosis |

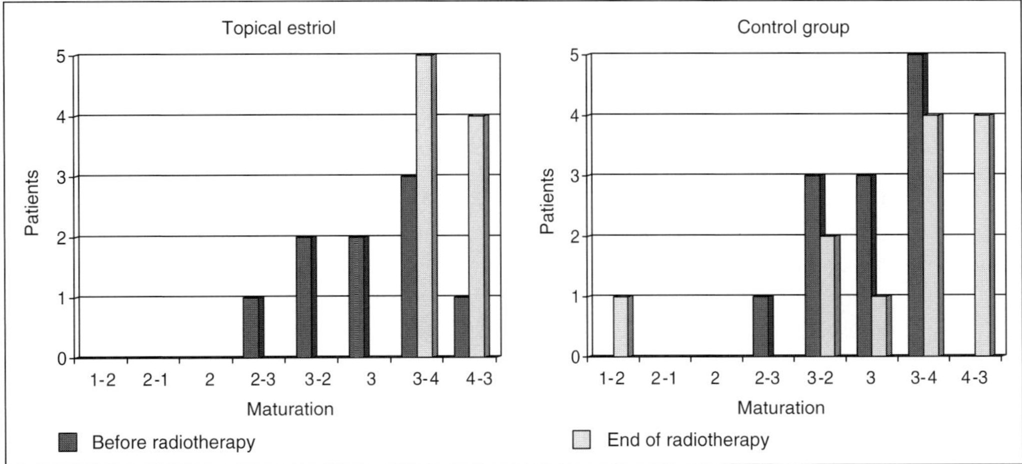


**_Fig. 2._** Change of maturation over treatment time.

with vaginal brachytherapy only (2/5) and about 23% (3/13) of the patients with cervix carcinoma. Within the estrogen group 4/9 patients (44%) showed an acute reaction, in the control group 4/12 (33%).

At the beginning of radiotherapy most premenopausal women exhibited a high maturation grade of 3–4. In the postmenopausal patients maturation ranged between 2–3 and 3–4. During treatment maturation increased or stayed at a high level in all patients of the estrogen group. In the control group there was no uniform change of the maturation level during radiotherapy (fig. 2): In 3 patients a decrease, in 3 patients no change, and in 6 patients an increase in maturation was observed. This was independent of age, menopausal status and treatment schedule.

The 18 postirradiation smears taken about 6 weeks after the end of therapy showed, with the exception of 1 with grade 1–2, mostly moderate to high levels of maturation (2–3 to 3–4). During the course of radiotherapy a maturation level ⩾3 was associated with an acute clinical reaction in 10/76 (13%) smears. A maturation level <3 was associated with vaginitis in 7/17 smears (41%) (p = 0.01).

The bacterial flora changed during treatment. Before the onset of radiation lactobacilli were predominant in 6/21 patients. In the course of radiotherapy this rate decreased to 10/93. Predominantly fecal-type (coccoid) bacteria were seen in all patients with vaginitis (p = 0.19) and lactobacilli dominated only in maturation levels ⩾3. The cellular radiation damage increased during the treatment time. At the end it was low in 4 patients, moderate in 9 and high in 8 patients. In our patients there was no correlation between the degree of cell changes and cytohormonal status or mucosal reaction.


## Discussion

Acute radiation vaginitis can be a major problem in radiotherapy of gynecological tumors. Mucosal reactions – ranging gradually from erythema with edema, circumscript or confluent mucositis to ulceration and necrosis – cause severe discomfort with pain and itching. Bacterial or fungal superinfections are favored by the damaged epithelium and the altered vaginal microenvironment and can aggravate the clinical state [9, 26]. Radiation vaginitis can necessitate interruption of the treatment protocol and thus result in decreased tumor control. Reports on acute side effects usually focus on reactions of the rectum or urinary bladder. Hence, information about radiation vaginitis in the literature is scarce.

In our study we could demonstrate a correlation between the rate of vaginitis and treatment modality, which are associated with different radiation doses at the mucosa. All patients treated with vaginal brachytherapy plus teletherapy developed acute mucosal reactions, in contrast to 40% of the patients who had vaginal brachytherapy only, and to about 23% of the patients with combined radiotherapy (without vaginal brachytherapy) for cervical cancer. Similar data were reported by Kucera et al. [19], who found a vaginitis rate of 49% in postoperative vaginal $^{192}$Ir high-dose-rate brachytherapy, when delivering 2 fractions of 10 Gy (2 cm from axis) and of 23% after 2 fractions of 7 Gy.

We could also detect a correlation between the cytohormonal status and radiation vaginitis. In high maturation levels ⩾3 vaginitis was diagnosed in 13% of the smears in contrast to 41% in low maturation levels <3. The estrogen effect, with a thick epithelium, is obviously associated with a decreased radiation effect, as reflected by the clinical results.

The estrogen effect also influences radiation-induced vaginitis by causing an antibacterial microenvironment and thus preventing superinfection. In high (estrogenic) maturation levels the flora is dominated by Doederlein's lactobacilli which create an acid antibacterial pH [4, 22]. This mechanism is destroyed by estrogen deficiency and by other factors like systemic antibiotics or alcaline tumor exudate, so that bacteria of fecal type can dominate [2, 10, 21]. In our patients the microenvironment changed during the course of radiotherapy: before the onset of irradiation an intact Doederlein flora was found in about 29% of the women in contrast to about 10% at the end of radiotherapy. Because in the Doederlein system, estrogen is required for fermentation of glycogen, lactobacilli were only predominant in maturation levels $\geq 3$. All patients with vaginitis showed predominantly bacteria of fecal type, which indicates the protective effect of the estrogen-dependent Doederlein flora.

Despite the influence of the cytohormonal status on radiation vaginitis, we did not observe a correlation between age or menopause status and acute reactions. In premenopausal patients the vaginitis rate was lower than in postmenopausal patients. However, they all were in the lowest dose group. This corresponds with the fact that there is not the typical age-related maturation pattern in patients with carcinoma of the uterus [5, 8, 28]. Patients with endometrial cancer have a higher mucosal maturation than patients with cervix cancer and in general these have a higher maturation than women without a tumor. This difference is seen in all age groups but is most marked in the postmenopausal patients. Cassano et al. [5] compared the cytohormonal status of 100 postmenopausal women with cancer with an age-matched control group without tumor. High maturation was noted in 69% of the patients with endometrial adenocarcinoma compared to 19% of the control, and in 46% of the patients with cervix carcinoma compared to 11% of the control. None of the index cases were atrophic, in contrast to 31% in the control group.

Similarly, all patients in our study showed a moderate or high maturation level at the beginning of radiotherapy and there was no patient with atrophic epithelium. During the course of radiotherapy the maturation pattern changed. In all patients, using estriol ovula, the level of maturation increased or stayed high. This indicates that the estrogen effect is functional during fractionated irradiation. In the control group the change of maturation did not follow a systematic pattern. In some patients, maturation increased, in some it decreased and some did not show a change. This was independent of menopause, age and treatment protocol. It remains unclear whether this is an effect due to proliferation, which is altered by radiotherapy, or a physiological reaction to changes of peripheral estrogen levels. More patients and a longer follow-up are required to answer this question.

There are no data about the cytohormonal level during radiotherapy but in some studies postirradiation changes of the maturation pattern are reported [1, 18, 24, 30]. Mostly a predominance of parabasal cells, reflecting low maturation levels, mixed with inflammatory cells, was found in smears taken about

6 months to 25 years after radiotherapy [1, 18]. Pitkin and Bradbury [24] assessed vaginal smears at intervals of 1 day to 3 years after completion of radiation. The thickness and maturation of the epithelium increased progressively with time. They concluded that a loss of virtually all epithelium was present in the areas receiving the maximal radiation dose, and that this denudation persisted for the first 3–6 months after radiotherapy. In our first follow-up examinations, about 6 weeks after the end of irradiation, most patients showed moderate or high maturation levels. The examined mucosa, however, received lower doses.

So far we have not been able to detect a decrease of the vaginitis rate in patients applying estrogen during radiotherapy. This result may be attributed to the small number and the inhomogeneity of patients in the two groups. There are no data about the effects of topical estrogen application during radiotherapy. In some studies the use of local estrogen after irradiation and its influence on late radiation injury was investigated [1, 18, 24, 25]. In 1965 it was reported that the topical use of estrogen starting at any time after completion of irradiation was associated with 'dramatic restauration' of relatively normal cytological morphology [24]. In a controlled double-blind study, patients showed a significant improvement in the vaginal epithelium, the gross appearance of the vagina (vaginal caliber, adhesions) and the clinical symptoms (dyspareunia). It was concluded that the application of estrogen after radiotherapy could lower the incidence of vaginal complications postirradiation [25].

Vaginal smear cytology is frequently used as a means of detection of the recurrence or persistence of malignant tumor cells in the follow-up of patients irradiated for genital neoplasms [18, 27, 29, 30]. Reliability is accepted despite difficulties in distinguishing preneoplastic and cancerous changes from benign radiation changes [7, 30]. In a study in postirradiation PAP smears Rintala et al. [27] had detected radiation-induced atypia in 28% of the vaginal smears taken during the first 4 months after radiotherapy, with a decreasing rate thereafter. Kaufman et al. [18] observed cellular radiation effects in 72% of the cases, as long as 25 years after irradiation. Few studies focus on cellular changes in the vaginal smear during radiotherapy. In benign and malignant cells alterations, regarding cell size, vacuolation of cytoplasm, multinucleation and nuclear changes are described [14, 23]. Already in 1947 Graham [11, 12] made similar observations in vaginal smears and drew prognostic conclusions for tumor persistence or recurrence. In our study the degree of cellular radiation changes increased in the individual towards the end of radiotherapy, but there was no correlation with other parameters such as treatment dose, cytohormonal status or vaginitis. This corresponds to observations by Zimmer [30] who showed that late radiation effects varied individually and were independent of dose and irradiation technique, patient's age and clinical and pathological findings. The variation did not seem to have any prognostic significance.

## Conclusion

Women with a low vaginal epithelial maturation level have a higher risk for developing acute radiation vaginitis compared to women with a high maturation level. These patients can be identified by a vaginal smear taken several days before the onset of radiotherapy, and then can receive intensified supportive care during treatment, which should stimulate the maturation of epithelium and improve the vaginal microenvironment. The evaluation of further patients is necessary in order to demonstrate whether topical estrogen, which shows both effects, also has a prophylactic potential for acute radiation vaginitis.

## References


1   Abitbol MM, Davenport JH: The irradiated vagina. Obstet Gynecol 1974;44:249–256.
2   Baron JA, Adams P, Ward M: Cigarette smoking and other correlates of cytologic estrogen effect in postmenopausal women. Fertil Steril 1988;50:766–771.
3   Blair OM: Hormonal cytopathology of the vagina; in Gold JJ, Josimovich JB (eds): Gynecologic Endocrinology. New York, Plenum Press, 1987, pp 159–199.
4   Bygdeman M, Swahn ML: Replens versus dienoestrol cream in the symptomatic treatment of vaginal atrophy in postmenopausal women. Maturitas 1996;23:259–263.
5   Cassano PA, Saigo PE, Hajdu SI: Comparison of cytohormonal status of postmenopausal women with cancer to age-matched controls. Acta Cytol 1986;30:93–98.
6   Cox JD, Stetz JA, Pajak TF: Toxicity criteria of the Radiation Therapy Oncology Group (RTOG) and the European Organization for Research and Treatment of Cancer (EORTC). Int J Radiat Oncol Biol Phys 1995;31:1341–1346.
7   Davey DD, Zaleski S, Sattich M, et al: Prognostic significance of DNA cytometry of postirradiation cervicovaginal smears. Cancer 1998;84:11–16.
8   Efstratiades M, Tamvakopoulou E, Papatheodorou B, et al: Postmenopausal vaginal cytohormonal pattern in 597 healthy women and 301 patients with genital cancer. Acta Cytol 1982;26:126–130.
9   Fraunholz IB, Schopohl B, Böttcher HD: Management of radiation injuries of vulva and vagina. Strahlenther Onkol 1998;174(suppl III):90–92.
10  Gordon AN, Martens M, La Pread Y, et al: Response of lower genital tract flora to external pelvic irradiation. Gynecol Oncol 1989;35:233–235.
11  Graham RM: Effect of radiation on vaginal cells in cervical carcinoma: Description of cellular changes. Surg Gynecol Obstet 1947;84:153–165.
12  Graham RM: Effect of radiation on vaginal cells in cervical carcinoma: Prognostic significance. Surg Gynecol Obstet 1947;84:166–173.
13  Grigsby PW, Russel A, Bruner D, et al: Late injury of cancer therapy on female reproductive tract. Int J Radiat Oncol Biol Phys 1995;31:1281–1299.
14  Gupta S, Mukherjee K, Gupta YN, et al: Sequential changes in cytology of vaginal smears in carcinoma of cervix uteri during radiotherapy. Int J Gynaecol Obstet 1987;25:303–308.
15  Himmelmann A, Notter G, Turesson I: Normal tissue reaction to high dose-rate intracavitary irradiation of the vagina with different fraction schedules and dose levels. Strahlentherapie 1985;161:163–167.
16  Hintz BL, Kagan AR, Chan P, et al: Radiation tolerance of the vaginal mucosa. Int J Radiat Oncol Biol Phys 1980;6:711–716.
17  Johnson RJ, Carrington BM: Pelvic radiation disease. Clin Radiol 1992;45:4–12.
18  Kaufman RH, Topek NH, Wall JA: Late irradiation changes in vaginal cytology. Am J Obstet Gynecol 1961;81:859–865.

19   Kucera H, Ünel N, Weghaupt K: Nebenwirkungen der postoperativen High-dose-rate-Iridium- und Low-dose-rate-Radium-Bestrahlung beim Korpuskarzinom. Strahlenther Onkol 1986;162: 111–114.

20   Maciejewski B, Withers HR, Taylor JM, et al: Dose fractionation and regeneration in radiotherapy for cancer of the oral cavity and oropharynx. 2. Normal tissue responses: Acute and late effects. Int J Radiat Oncol Biol Phys 1990;18:101–111.

21   Milsom I, Arvidsson L, Ekelund P, et al: Factors influencing vaginal cytology, pH and bacterial flora in elderly women. Acta Obstet Gynecol Scand 1993;72:286–291.

22   Nachtigall LE: Comparative study: Replens versus local estrogen in menopausal women. Fertil Steril 1994;61:178–180.

23   Picha E: Die zytologischen Veränderungen im Vaginalsmear während der Strahlenbehandlung des Carcinoma colli uteri. Krebsarzt 1956;11:22–28.

24   Pitkin RM, Bradbury JT: The effect of topical estrogen on irradiated vaginal epithelium. Am J Obstet Gynecol 1965;92:175–182.

25   Pitkin RM, Van Voorhis LW: Postirradiation vaginitis. An evaluation of prophylaxis with topical estrogen. Radiology 1971;99:417–421.

26   Rauthe G: Management of reactions and complications following radiation therapy; in Vahrson HW (ed): Radiation Oncology of Gynecological Cancers. Berlin, Springer, 1997, pp 433–454.

27   Rintala MA, Rantanen VT, Salmi TA, et al: PAP smear after radiation therapy for cervical carcinoma. Anticancer Res 1997;17:3747–3750.

28   Rubio CA, Hvalec S, Pareja A: The value of cytohormonal studies (karyopyknotic index) for detecting recurrence of carcinoma. Acta Cytol 1967;11:176–178.

29   Shield PW: Chronic radiation effects: A correlative study of smears and biopsies from the cervix and vagina. Diagn Cytopathol 1995;13:107–119.

30   Zimmer TS: Late irradiation changes. A cytological study of cervical and vaginal smears. Cancer 1959;12:193–196.

Ingeborg B. Fraunholz, Department of Radiation Oncology,
Johann Wolfgang Goethe University, Theodor-Stern-Kai 7,
D–60590 Frankfurt am Main (Germany)
Tel. +49 69 6301 5130, Fax +49 69 6301 5091, E-Mail fraunholz@vff.uni-frankfurt.de

Dörr W, Engenhart-Cabillic R, Zimmermann JS (eds): Normal Tissue Reactions in Radiotherapy and Oncology. Front Radiat Ther Oncol. Basel, Karger, 2002, vol 37, pp 121–127

# A Study of High Frequency Ultrasound to Assess Cutaneous Oedema in Conservatively Managed Breast

*C. Wratten, J. Kilmurray, S. Wright, P. O'Brien, M. Back, C. Hamilton, J. Denham*

Department of Radiation Oncology, Mater Misericordiae Hospital, Newcastle, Australia

Analysis of the factors predisposing to acute normal tissue toxicity in patients treated conservatively for breast cancer has been limited by the lack of reliable quantitative measures. Most studies have used subjective semiquantitative measures, which are subject to significant inter- and intra-observer variation. This variation limits the interpretation of any results obtained. As we were interested in studying the factors influencing skin reactions in patients receiving adjuvant whole breast irradiation, we were interested in assessing quantitative measures.

High frequency ultrasound has been used in dermatological practice for several years and has been shown to be an accurate method of assessing changes in skin thickness [4–6, 11]. Three studies [7, 8, 12] have suggested that radiotherapy may be associated with increased skin thickness and two of these studies have involved examination of the conserved breast [8, 12]. In view of these reported findings we undertook a study to assess the usefulness of high frequency ultrasound as a quantitative measure of radiation skin reactions in the conserved breast. This presentation is an interim analysis of this study.

## Material and Methods

Thirteen patients who received whole breast irradiation form the basis of the study. All patients had undergone wide local excision plus or minus axillary dissection prior to referral for radiotherapy. Patients were asked to participate in the study at the time of referral for

***Table 1.*** Patient characteristics

| Patient | Age years | Stage | Axillary surgery | Nodal involvement | Clinical oedema |
|---|---|---|---|---|---|
| 1 | 54 | T1N0 | Level 2 | 0/12 | No |
| 2 | 54 | T1N1 | Level 2 | 1/11 | Yes |
| 3 | 69 | T1Nx | Nil | – | No |
| 4 | 54 | T2N0 | Sentinel | 0/3 | Yes |
| 5 | 47 | T2N0 | Sentinel | 0/3 | No |
| 6 | 56 | T1Nx | Nil | – | No |
| 7 | 68 | TisNx | Nil | – | No |
| 8 | 48 | T1N0 | Sentinel | 0/1 | Yes |
| 9 | 58 | TisNx | Nil | – | No |
| 10 | 53 | T1N0 | Level 2 | 0/14 | No |
| 11 | 77 | T1Nx | Nil | – | No |
| 12 | 43 | T2N0 | Level 2 | 0/24 | Yes |
| 13 | 60 | T1N0 | Level 2 | 0/14 | Yes |

radiotherapy. Written informed consent was obtained from all patients. All patients were treated with tangential megavoltage photon fields (4–18 MV) plus or minus a direct electron boost to the excision site. Total doses ranged between 50 and 64 Gy. Fraction size was 2 Gy in all patients. Patients were treated once daily, 5 days per week. One patient had received anthracycline-based chemotherapy prior to commencing radiotherapy. All but one patient were taking tamoxifen 20 mg daily.

Ultrasound measures were performed using a commercial 20 MHz high frequency ultrasound (Dermascan C, Cortex Technology, Denmark) with a medium focus transducer. This unit gives an axial resolution of 60 μm and a lateral resolution of 200 μm. The viewing field is up to 15 mm in depth. It is capable of both A and B mode scanning. B mode scans were used to assess skin thickness in this group of patients. The ultrasound unit has in-built software to allow determination of average tissue depth in the recorded image. Gain can be adjusted to overcome the reduction in reflected echogenicity with increasing depth or tissue density. Scans were obtained 4 cm medial and lateral to the nipple in both the treated and untreated breast. A predefined technique was used to minimize variation due to technical factors [13]. All patients were examined in supine position. Measurements were obtained prior to commencing radiotherapy and prior to fractions 4, 6, 9, 11, 16, 21 and 26.


## Results

The characteristics of the patients in this study are shown in table 1. Eight of the 13 patients had undergone axillary surgery. There was obvious visible breast oedema prior to commencing radiotherapy evident in 5 patients. Figure 1 shows the mean cutaneous breast thickness 4 cm medial and lateral to the nipple

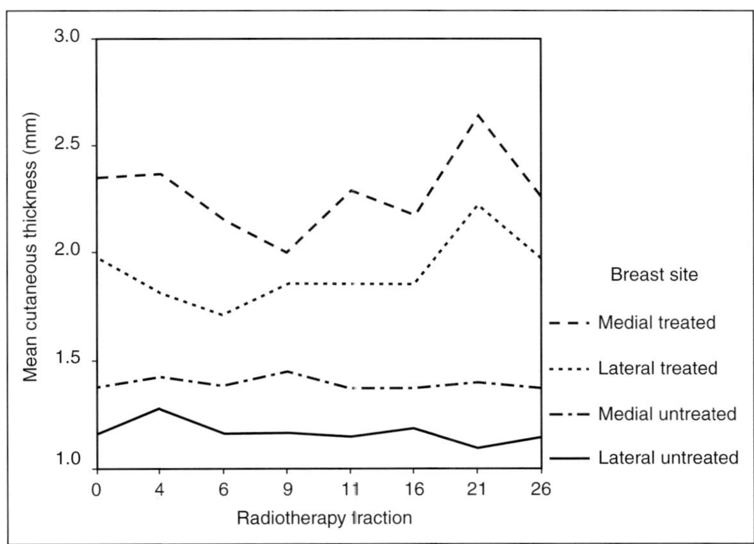


***Fig. 1.*** Mean breast skin thickness.

***Table 2.*** Variation in ultrasound-determined breast skin thickness

|  | Medial treated breast | Lateral treated breast | Medial untreated breast | Lateral untreated breast |
|---|---|---|---|---|
| Mean thickness, mm | 2.23 | 1.91 | 1.38 | 1.16 |
| Standard deviation, mm | 1.09 | 0.71 | 0.26 | 0.22 |
| Interpatient coefficient of variation, % | 48.9 | 37.23 | 18.9 | 18.9 |
| Intrapatient coefficient of variation, % | 15.8 | 14.4 | 9.4 | 8.8 |

in the treated and untreated breast of all patients. It is obvious, when considering the overall group of patients, that the treated breast skin is thicker than the untreated breast, and that this thickening is evident before the start of radiotherapy. In both the treated and untreated breast the medial aspect is thicker than the outer aspect. Analysis of the mean skin thickness of all patients during radiotherapy shows no obvious effect of radiation on breast skin thickness during the treatment course. These results are shown in tabular form in table 2. Interpatient and intrapatient coefficients of variation are minimal in the untreated breast. Within the treated breast they are both more substantial.

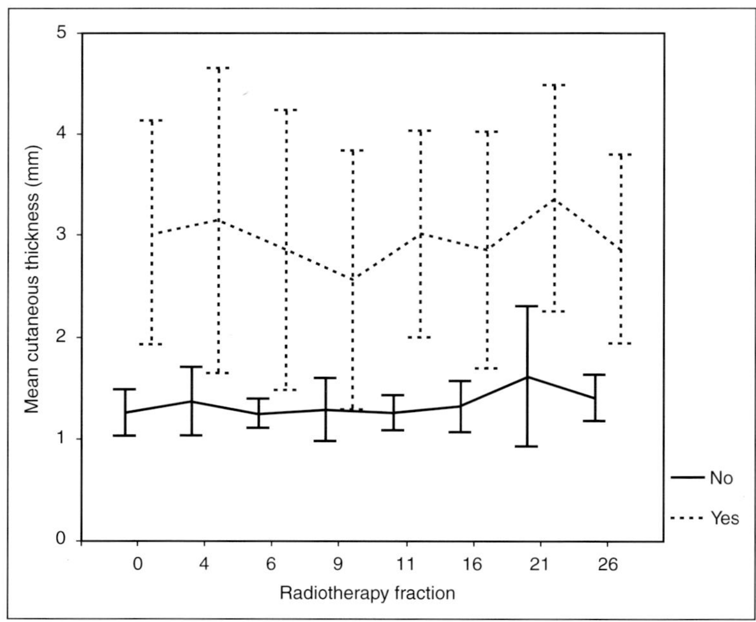


*Fig. 2.* Effect of axillary dissection on breast skin thickness: medial treated breast.

When patients were divided into those who had or had not had axillary surgery there was a marked difference in skin thickness (fig. 2, 3). This was most evident in the medial treated breast (fig. 2). Dividing patients into groups based on the presence or absence of visual breast oedema also resulted in marked observed differences in cutaneous thickness (fig. 4, 5).

## Discussion

At present we have assessed a limited number of patients and therefore no final conclusions can be gained from this data. It does appear, however, that high frequency ultrasound is a useful quantitative measure of cutaneous breast oedema. Measures obtained from the untreated breast show little inter- or intrapatient variability. The inter- and intrapatient coefficient of variation in the treated breast thickness is much greater. The reasons for these changes cannot be determined from such a small study. It is obvious, however, that axillary surgery does play an important role. Whether radiotherapy contributes to the alterations in skin thickness observed in this study will only become evident with a larger study. It is possible that radiotherapy may only induce changes in skin

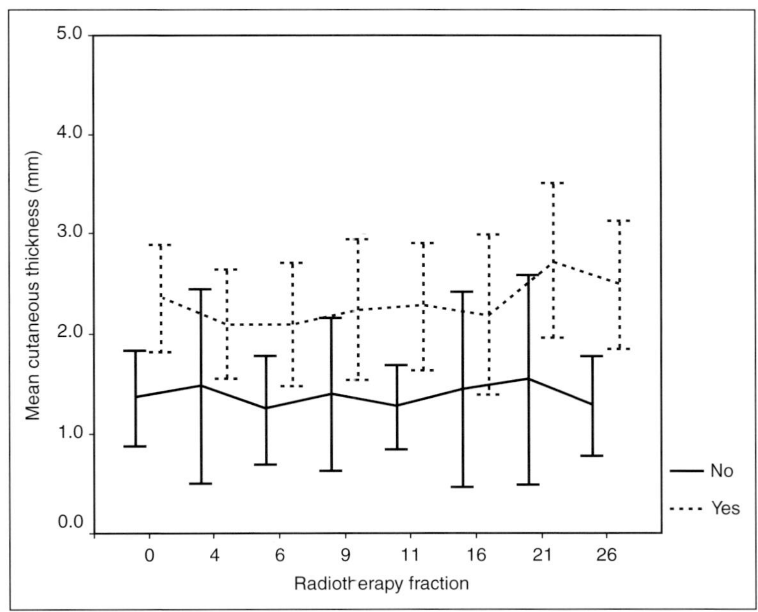

**Fig. 3.** Effect of axillary dissection on breast skin thickness: lateral treated breast.

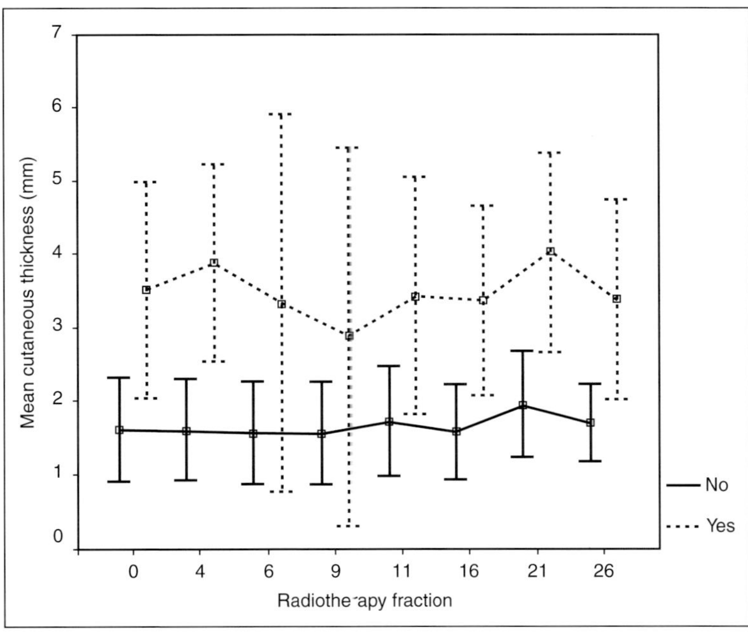

**Fig. 4.** Effect of visible breast oedema on breast skin thickness: medial treated breast.

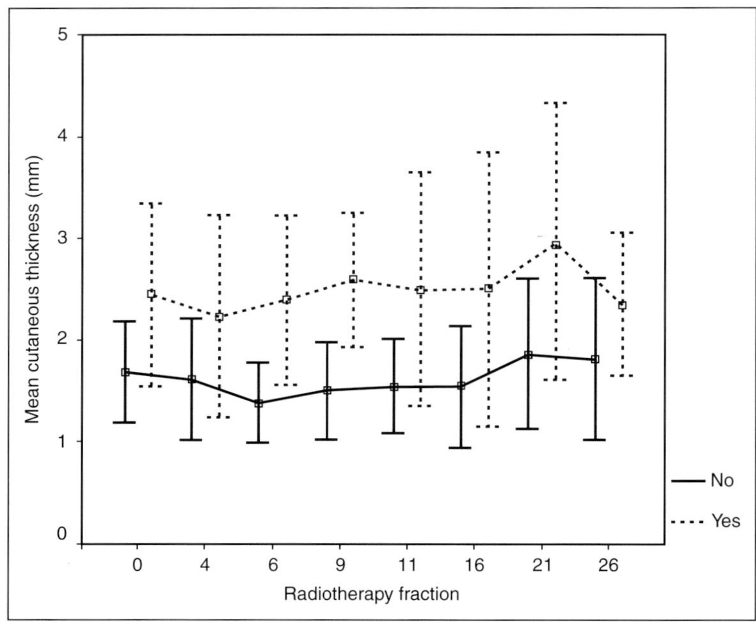


*Fig. 5.* Effect of visible breast oedema on breast skin thickness: lateral treated breast.

thickness in those patients who do not show an alteration in skin thickness prior to commencing radiotherapy. In our patients these are predominantly those who have not undergone an axillary dissection.

It also appears that breast skin thickness varies across the breast. Both in the treated and untreated breast the medial aspect was thicker than the lateral aspect. This appeared to be related to the presence or absence of axillary surgery and may therefore reflect differences in lymphatic drainage between the inner and outer aspects of the breast skin.

The patients with the most marked cutaneous breast thickness in our study were those with obvious visible breast oedema prior to commencing radiotherapy. Breast oedema is an unfortunate complication of the conservative management of breast cancer. It occurs in 9–39% of patients [1–3, 9, 10] and can result in considerable discomfort and poor cosmesis. Studies of breast oedema have also been limited by the lack of reliable quantitative measures. It is possible that cutaneous oedema measured via high frequency ultrasound may be a useful measure of breast oedema and allow the more accurate study of this complication. We intend to study this further.

Previous studies [8, 12] using high frequency ultrasound have also reported increased skin thickness in patients who have undergone breast-conserving therapy. These studies have not assessed the effect of axillary surgery on the results they report. The majority of patients reported in these studies have had measures performed only after commencing radiotherapy, therefore the effect of prior surgery has not been noted. We are continuing to enter patients into this study so that we can assess the effect of radiotherapy on breast skin thickness. To date, however, it does not appear that high frequency ultrasound is an ideal sensitive quantitative measure of acute radiation breast skin reactions in this group of patients.

## References


1 Beadle GF, Silver B, Botnick L, Hellman S, Harris JR: Cosmetic results following primary radiation therapy for early breast cancer. Cancer 1984;54:2911–2918.

2 Boyages J, Barraclough B, Middledorp J, Gorman D, Langlands AO: Early breast cancer: Cosmetic and functional results after treatment by conservative techniques. Aust NZ J Surg 1988; 58: 111–121.

3 Clarke D, Martinez A, Cox RS, Goffinet DR: Breast edema following staging axillary node dissection in patients with breast carcinoma treated by radical radiotherapy. Cancer 1982; 49:2295–2299.

4 Gniadecka M: Dermal oedema in lipodermatosclerosis: Distribution, effects of posture and compressive therapy evaluated by high-frequency ultrasonography. Acta Derm Venereol 1995; 75:120–124.

5 Gniadecka M, Jemec GB: Quantitative evaluation of chronological ageing and photoageing in vivo: Studies on skin echogenicity and thickness. Br J Dermatol 1998;139:815–821.

6 Gniadecka M, Serup J, Sondergaard J: Age-related diurnal changes of dermal oedema: Evaluation by high-frequency ultrasound. Br J Dermatol 1994;131:849–855.

7 Gottlöber P, Kerscher M, Korting H, Peter R: Sonographic determination of cutaneous and subcutaneous fibrosis after accidental exposure to ionising radiation in the course of the Chernobyl nuclear power plant accident. Ultrasound Med Biol 1997;23:9–13.

8 Le Floch O, Gonzague L, Feil-Bastid C, Aget H, Garaud P, Bougnoux P: Factors influencing long term modification of the irradiated skin thickness in patients conservatively treated for breast carcinoma. Radiother Oncol 1998;48(suppl 1):S108.

9 Pezner RD, Patterson MP, Hill LR, Desai KR, Vora N, Lipsett JA: Breast edema in patients treated conservatively for stage I and II breast cancer. Int J Radiat Oncol Biol Phys 1985;11:1765–1768.

10 Read PE, Ash DV, Thorogood J, Benson EA: Short term morbidity and cosmesis following lumpectomy and radical radiotherapy for operable breast cancer. Clin Radiol 1987;38:371–373.

11 Seidenari S, Pagnoni A, Di Nardo A, Giannetti A: Echographic evaluation with image analysis of normal skin: Variations according to age and sex. Skin Pharmacol 1994;7:201–209.

12 Warszawski A, Rottinger EM, Vogel R, Warszawski N: 20 MHz ultrasonic imaging for quantitative assessment and documentation of early and late postradiation skin reactions in breast cancer patients. Radiother Oncol 1998;47:241–247.

13 Wratten C, Kilmurray J, Wright S. et al: Pilot study of high frequency ultrasound to assess cutaneous oedema in the conservatively managed breast. Int J Cancer 2000;90:295–301.



C. Wratten, Department of Radiation Oncology, Newcastle Mater Misericordiae Hospital, Waratah, NSW 2298 (Australia)
Tel. +61 2 4921 1179, Fax +61 2 4960 2566, E-Mail mdcwr@alinga.newcastle.edu.au

Dörr W, Engenhart-Cabillic R, Zimmermann JS (eds): Normal Tissue Reactions in Radiotherapy and Oncology. Front Radiat Ther Oncol. Basel, Karger, 2002, vol 37, pp 128–131

# Intensity-Modulated Radiotherapy Reduces Lung Irradiation in Patients with Carcinoma of the Oesophagus

*Christopher M. Nutting*[a]*, James L. Bedford*[b]*, Vivian P. Cosgrove*[b]*, Diana M. Tait*[a]*, David P. Dearnaley*[a]*, Steve Webb*[b]

[a] Academic Department of Radiotherapy and [b] Joint Department of Physics, Institute of Cancer Research and Royal Marsden NHS Trust, Sutton, UK

The 2-year survival rate after conventional radiotherapy for carcinoma of the oesophagus is around 10–20% [8]. Concomitant chemoradiation schedules have produced survival figures of 25–30% at 5 years, and this is now considered standard treatment [1]. Conformal radiotherapy techniques offer the potential to deliver higher doses of radiation to oesophageal tumours [5], and this may improve local tumour control. However, concerns regarding late normal tissue damage to the lung parenchyma and spinal cord remain a concern. Intensity-modulated radiotherapy (IMRT) allows complex dose distributions to be produced, and can reduce the dose to radiosensitive organs close to the tumour [2]. The present study was designed to investigate the impact of beam intensity modulation on treatment planning for carcinoma of the oesophagus, by comparing a standard three-dimensional conformal radiotherapy (3DCRT) technique to an IMRT technique using the same number and orientation of treatment fields.

## Materials and Methods

*Patients and Radiotherapy Planning*
Five patients with oesophageal carcinoma recently treated were identified. The clinical target volume included the primary tumour with a circumferential margin of 2 cm, and a craniocaudal margin of 5 cm. A three-dimensional margin of 15 mm was added to the clinical target volume to account for movement, creating the planning target volume (PTV). Spinal cord and lung parenchyma were also outlined.

***Table 1.*** Mean results ($\pm$1SD) for 3DCRT and five-field IMRT in patients with carcinoma of the oesophagus

|  |  | 3DCRT | 5-field IMRT | Statistical significance |
|---|---|---|---|---|
| PTV | Mean dose, Gy | 55.7 ($\pm$1.0) | 55.7 ($\pm$1.0) | – |
|  | Minimum dose, % | 92.4 ($\pm$3.1) | 94.0 ($\pm$1.9) | p = 0.36 |
|  | Maximum dose, % | 108.4 ($\pm$4.2) | 105.8 ($\pm$1.8) | p = 0.18 |
|  | Dose range, % | 16.0 ($\pm$4.9) | 11.8 ($\pm$3.3) | p = 0.03 |
| Spinal cord | Maximum dose, Gy | 44.5 ($\pm$0.5) | 44.5 ($\pm$0.6) | p = 0.91 |
| Lungs | Mean dose, Gy | 11.0 ($\pm$2.9) | 9.5 ($\pm$2.3) | p = 0.001 |
|  | Volume receiving >18 Gy, % | 18.8 ($\pm$11.9) | 14.1 ($\pm$10.1) | p = 0.001 |
|  | NTCP, % | 1.0 ($\pm$0.7) | 0.6 ($\pm$0.4) | p = 0.008 |

*3DCRT and IMRT Planning*

A 3DCRT plan was created for each patient, using 6 MV two-phase technique. The first phase used parallel-opposed antero-posterior and postero-anterior fields and the second phase an anterior and two posterior wedged oblique fields at gantry angles of approximately 110° and 250°. All fields were conformally shaped using the beam's eye view. Beam weights were optimised to minimise PTV dose inhomogeneity. The maximum spinal cord dose remained below 45 Gy, and the irradiated volume of lung was minimised. IMRT plans using the same gantry angles were produced using CORVUS, an inverse treatment planning system [9]. The following constraints were used: PTV: goal dose 55 Gy ($\pm$5%) in 30 fractions; lungs: 18 Gy to less than 5–10% of the lung volume; spinal cord: maximum dose limit 45 Gy.

*Comparison of Treatment Plans*

The mean PTV dose, PTV dose range, spinal cord maximum dose, and the mean lung dose were recorded for each plan. The mean lung dose was used as a surrogate endpoint for radiation pneumonitis, as this correlates closely to clinical reports of pneumonitis [6]. Normal tissue complication probability (NTCP) for lung was calculated using the parameters proposed by Kwa et al. [6]. Statistical significance of each comparison was assessed using a two-tailed Student's t test.

## Results

The mean minimum and maximum doses to the PTV for the 3DCRT plans and the IMRT plans were well within the constraints of 90–110 % (table 1). The mean PTV dose range (inhomogeneity) for the 3DCRT plans was significantly higher than for the IMRT plans (p = 0.03). The maximum spinal cord dose was less than the 45 Gy constraints for all plans, and the IMRT plans also reduced the volume of spinal cord that received lower doses of radiation in most cases. The mean lung dose with the IMRT plans was significantly reduced compared to the 3DCRT plans (p = 0.001). In all patients, the volume of lung receiving higher doses of radiation was reduced, although the absolute maximum dose was not reduced because the lung parenchyma is adjacent to the PTV. NTCP calculations showed a significant reduction in risk of grade $\geq 2$ radiation pneumonitis with 5-field IMRT (p = 0.008).

## Discussion

IMRT was associated with an improvement in the PTV dose homogeneity, and a reduction in the mean lung dose for equivalent maximum radiation doses to the spinal cord compared to 3DCRT. The reduction in mean lung dose was 1.5 Gy. The predicted benefit of IMRT for oesophageal carcinoma, where the PTV is cylindrical, is relatively small compared to other tumour sites where the PTV is concave [7].

Reductions of a similar magnitude to this have been reported by other authors who compared 3DCRT and noncoplanar IMRT for stage III non-small cell lung carcinoma [3] and this has been used as a basis for dose escalation of this tumour site [4]. The absolute NTCP values were lower than that seen clinically in this patient group. This difference is likely to be due to the use of concomitant chemotherapy, which has not been accounted for in the NTCP parameters. The IMRT technique was associated with a 40% reduction in the NTCP parameter predicting radiation pneumonitis of at least grade 2. Although the NTCP is recognised to be only an estimate of lung damage, the calculated probabilities are consistent with the dosimetric statistics.

## Conclusions

The comparisons between 3DCRT and IMRT techniques demonstrated that IMRT reduces the mean lung dose, and improves PTV homogeneity. The clinical implementation of IMRT in patients with oesophageal carcinoma may cause less radiation pneumonitis, or may allow moderate degrees of dose escalation.

## References


1    Bosset JF, Gignoux M, Triboulet JP, et al: Chemoradiotherapy followed by surgery compared with surgery alone in squamous cell cancer of the oesophagus. N Engl J Med 1997;337:161–167.
2    Brahme A: Optimisation of stationary and moving beam radiation therapy techniques. Radiother Oncol 1988;12:129–140.

3   Derycke S, De Gersem WRT, Van Duyse BBR, De Neve WCJ: Conformal radiotherapy of stage III non-small cell lung cancer: A class solution involving non-coplanar intensity-modulated beams. Int J Radiat Oncol Biol Phys 1998;41:771–777.

4   Derycke S, Van Duyse B, De Wagter C, et al: Non-coplanar beam intensity modulation allows large dose escalation in stage III lung cancer. Radiother Oncol 1997;45:253–261.

5   Guzel Z, Bedford JL, Childs PJ, et al: A comparison of conventional and conformal radiotherapy of the oesophagus: Work in progress. Br J Radiol 1998;71:1076–1082.

6   Kwa SLS, Lebesque N, Theuws JCM, et al: Radiation pneumonitis as a function of mean lung dose: An analysis of pooled data of 540 patients. Int J Radiat Oncol Biol Phys 1998;42:1–9.

7   Nutting C, Rowbottom C, Convery DJ, et al: Intensity modulated radiotherapy in tumours of the head and neck. Br J Radiol 1999;72:98–99.

8   Smith TJ, Ryan LM, Douglass HO, et al: Combined chemo-radiotherapy vs. radiotherapy alone for early stage squamous cell carcinoma of the oesophagus: A study of the Eastern Co-operative Oncology Group. Int J Radiat Oncol Biol Phys 1998;42:269–276.

9   Xing L, Curran B, Hill R, et al: Dosimetric verification of a commercial inverse treatment planning system. Phys Med Biol 1999;44:463–478.

Dr. C. Nutting, FRCR, Academic Department of Radiotherapy and Oncology,
3rd floor Orchard House, Royal Marsden NHS Trust, Downs Road,
Sutton, Surrey SM2 5PT (UK)
Tel. +44 208 661 3271, Fax +44 208 643 8809, E-Mail chrisnutting@cs.com

Dörr W, Engenhart-Cabillic R, Zimmermann JS (eds): Normal Tissue Reactions in Radiotherapy and Oncology. Front Radiat Ther Oncol. Basel, Karger, 2002, vol 37, pp 132–139

# Impaired Sphincter Function and Good Quality of Life in Anal Carcinoma Patients after Radiotherapy: A Paradox?

*Dirk Vordermark*[a], *Marco Sailer*[b], *Michael Flentje*[a], *Arnulf Thiede*[b], *Oliver Kölbl*[a]

[a] Department of Radiation Oncology and
[b] Department of Surgery, University of Würzburg, Germany

Recent large trials have confirmed a combination of definitive radiotherapy and simultaneous chemotherapy with fluorouracil and mitomycin as the standard treatment for most cases of anal epidermoid carcinoma. Colostomy-free survival after such treatment was reported to range between 50 and 71% at 3 or 4 years [2, 5, 20]. However, several issues concerning the optimal treatment concept in this disease remain unresolved, among them total external beam dose, type of boost treatment (external beam, interstitial or intracavitary brachytherapy), indication for boost treatment (all patients or patients with insufficient response to initial external beam treatment only) and time interval between main series and boost treatment. Also, the advantage of chemoradiation over an initial surgical management has been questioned in the very recent surgical literature [18] and very low external beam doses have been recommended by some authors [7]. In such a patient cohort with a favorable prognosis, as compared with most other malignant diseases, quality of life (QoL) evaluation should have an important role in comparing different treatment strategies and must therefore be included in randomized trials.

Colostomy-free survival has been chosen as a main endpoint in most investigations of anal carcinoma, thus taking into account the preservation of anorectal function. In a recent study from our institution, QoL and actual sphincter function were evaluated in colostomy-free survivors from an overall cohort of 39 patients with anal carcinoma [21].

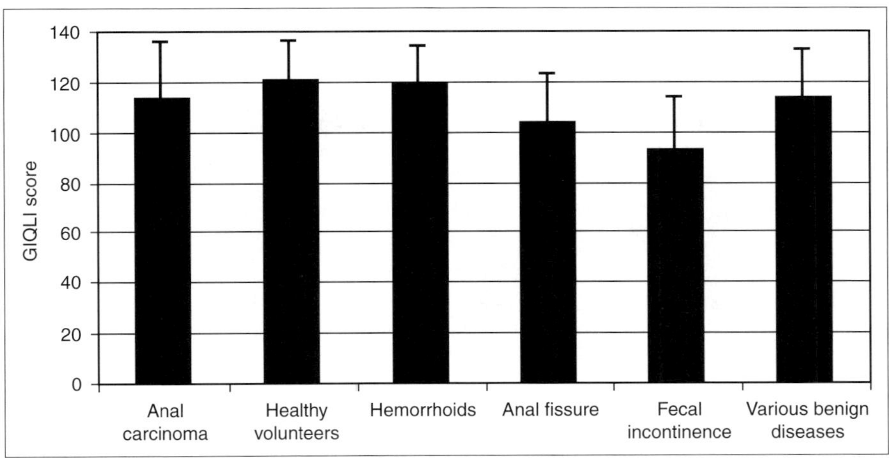



*Fig. 1.* GIQLI in 22 colostomy-free survivors 3.1 ± 3.1 years after radiotherapy or chemoradiation for anal carcinoma, as compared with published data for healthy volunteers and patients with benign anorectal diseases [13]. Values are given as mean ± standard deviation. The maximum GIQLI score, reflecting best QoL, is 144. No statistical comparison is made due to age differences.

QoL was measured using the validated Gastrointestinal Quality of Life Index (GIQLI) introduced by Eypasch et al. [4], consisting of 36 items relating to gastrointestinal disease, each scoring 0–4 points, leading to a maximum score (best QoL) of 144. The mean score in anal carcinoma patients was 114, a value similar to published results [13] of cohorts with less severe anorectal diseases or healthy volunteers (fig. 1). Although no formal statistical comparison was done due to age differences between groups, the fact that GIQLI scores decrease with older age and that the anal carcinoma group was by far the oldest led to the interpretation that QoL scores were unexpectedly high in the colostomy-free survivors.

Anorectal manometry performed in 16 patients from this group, however, revealed a significant reduction in sphincter length (extent of the high pressure zone), resting pressure and maximum squeeze pressure, as compared with normal volunteers (fig. 2).

The question arose why QoL scores in colostomy-free survivors were similar to those of healthy controls despite quite severe impairment of anorectal function. Therefore, in the present analysis, functional parameters, single symptoms and a continence score were correlated with the overall QoL value as measured by GIQLI to determine which parameters are most crucial for QoL in this group.

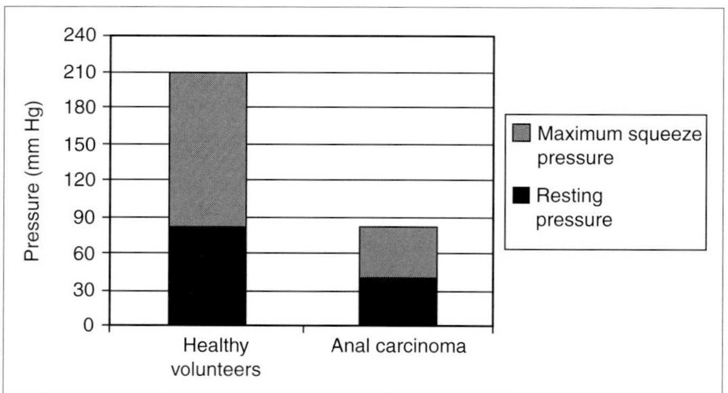


***Fig. 2.*** Comparison of median anorectal manometry results in healthy volunteers (n = 21) and anal carcinoma patients (n = 16) 2.5 ± 2.2 years after radiotherapy or chemoradiation. Both resting pressure and maximum squeeze pressure, the increase from baseline upon voluntary contraction, were significantly different between groups (p < 0.05, Mann-Whitney U test). A minimum of 60 mm Hg resting pressure plus another 60 mm Hg maximum squeeze pressure has been described as necessary for maintaining continence [3].

## Material and Methods

*Patients and Treatment*

All 16 patients in whom anorectal manometry data was available were included in the analysis. Of these, there were 13 women (81%), mean age (±standard deviation) was 63 ± 16 years. The tumor location was anal canal in 14 (87%) and margin in 2 (13%) patients. According to the International Union Against Cancer (UICC), T stages were T1 in 2 (13%), T2 in 12 (75%), and T3 and T4 each in 1 (6%) patient [19]. 12 (75%) cases were classified as N0. CT-based planning and photon treatment delivered by a linear accelerator were applied in all cases. Mean external beam dose at the ICRU reference point was 56.5 ± 4.4 Gy (range 50–66 Gy). 11 patients (69%), all with anal canal tumors, received an additional intracavitary boost treatment with iridium-192 according to a previously published method [8] in one or two sessions with resulting total boost doses at 5 mm depth of 5–10 Gy. Chemotherapy with two courses of fluorouracil and mitomycin was given to 13 of the patients (81%).

*QoL, Anorectal Manometry and Continence*

The QoL and manometry data used in the present analysis were acquired 2.5 ± 2.2 years after treatment, as previously described [21]. Briefly, patients were administered the 36-item GIQLI questionnaire and a separate, well-established 10-item continence questionnaire with a maximum total score of 36 points equaling complete continence.

Anorectal manometry was performed by automatic stepwise retraction of a fluid-perfused catheter for pressure recording at various levels of the continence organ. The measurements of spincter length (distance from start of pressure increase to point where pressure drops to 0 when probe leaves anal canal), resting pressure (highest pressure in anal canal when sphincter relaxed) and maximum squeeze pressure (highest rise from resting pressure baseline during voluntary contraction) were used in the present analysis.

*Table 1.* Correlation of QoL with functional and symptom parameters

| Parameter | r | p |
|---|---|---|
| Continence score | 0.71 | 0.002 |
| Sphincter length | −0.86 | 0.751 |
| Resting pressure | 0.38 | 0.140 |
| Maximum squeeze pressure | 0.49 | 0.052 |
| Rectal bleeding | 0.31 | 0.241 |
| Urgency of defecation | 0.57 | 0.020 |
| Diarrhea | 0.51 | 0.042 |
| Fatigue | 0.89 | <0.001 |

Analysis of correlation of functional and symptom parameters with GIQLI in 16 colostomy-free survivors treated with radiotherapy or chemoradiation for anal carcinoma, based on calculation of Spearman-rank correlation coefficient (r).

*Statistical Analysis*

Manometry results, continence score and several symptom scores derived from single items of the GIQLI questionnaire were entered into an analysis of correlation with the total GIQLI score. Specifically, the symptoms of rectal bleeding, urgency of defecation, diarrhea and fatigue, as classified by the patients, were analyzed. The Spearman-rank correlation coefficient r and the corresponding p value were calculated for each pair of data sets.

## Results

Calculation of the Spearman-rank correlation coefficient revealed a highly significant correlation of GIQLI score with the fecal continence score derived from a separate questionnaire (table 1). None of the anorectal manometry parameters was significantly correlated with GIQLI, although there was a trend toward an association with maximum squeeze pressure, the effect of voluntary contraction of the external sphincter. Of the symptoms analyzed, diarrhea, urgency of defecation and, in particular, fatigue but not rectal bleeding showed significant correlation with QoL as calculated by GIQLI. The distribution of individual values for selected parameters is illustrated in figure 3.

## Discussion

In a previous investigation, anorectal manometry data on 16 colostomy-free survivors irradiated for anal carcinoma was reported from our institution, in

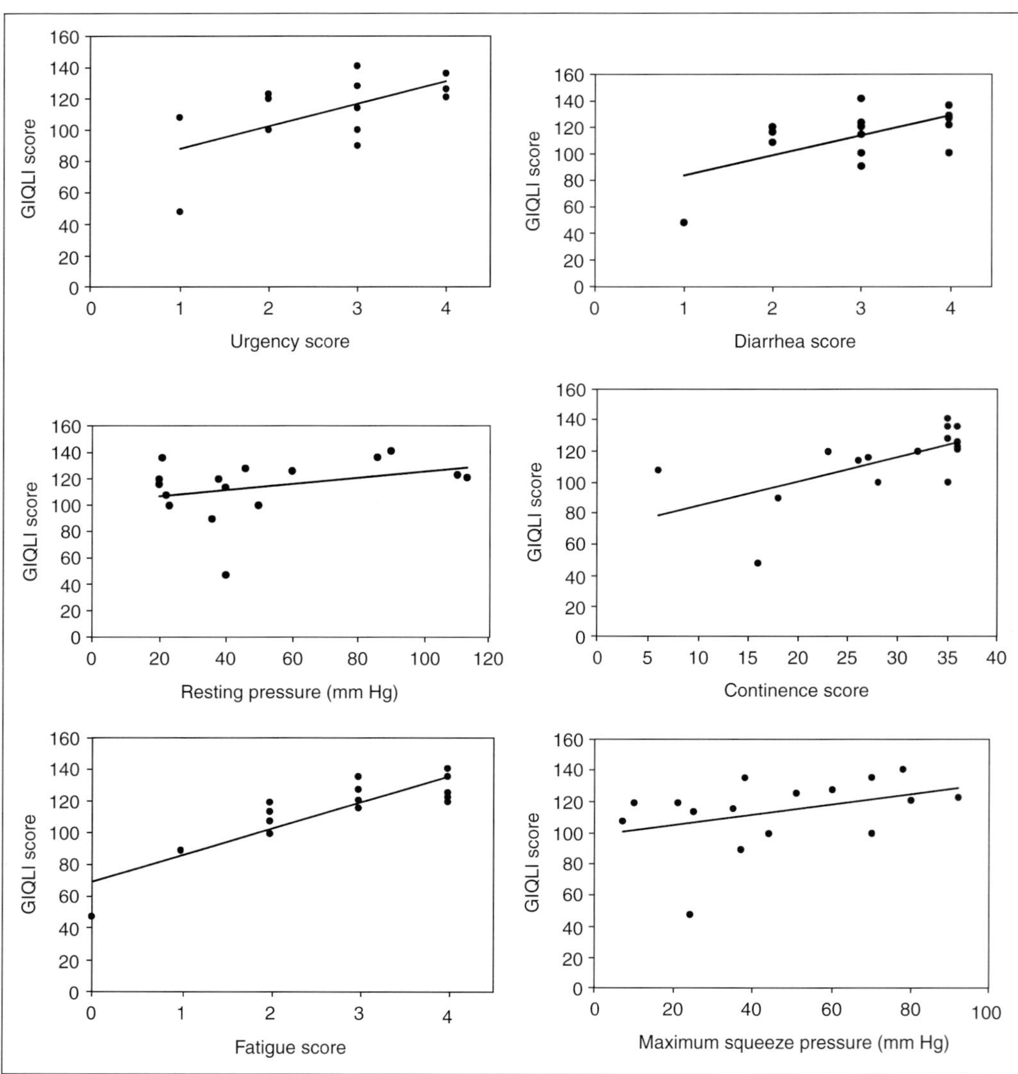



***Fig. 3.*** Correlation of functional and symptom parameters with QoL. GIQLI in 16 colostomy-free survivors treated with radiotherapy or chemoradiation for anal carcinoma. Symptom scores (right-hand side) were taken from single items of GIQLI, where low scores reflect severe impairment by symptoms, resulting in mainly positive correlations with QoL.

conjunction with QoL measurements [21]. Although still limited in size, this series represents, to the authors' knowledge, the largest such cohort for which both information is available. In an uncommon disease such as anal carcinoma, a thorough analysis of the importance of functional parameters and symptoms for QoL may be useful, in particular to explain the discrepancy between near-normal QoL scores and severely reduced anorectal pressure values in the original publication.

In the present analysis, continence as measured by a separate 10-item questionnaire was closely correlated with QoL expressed as GIQLI score. This finding reflects the importance of the quality of sphincter function in organ-conserving therapy: patients considered to be successfully treated because they are colostomy-free survivors may still suffer from severely reduced QoL due to impaired sphincter function. Of the three manometry parameters significantly affected in anal carcinoma patients – sphincter length, resting pressure and maximum squeeze pressure – none was significantly correlated with QoL although there was a trend toward a correlation for the latter. While a minimum resting pressure of 60 mm Hg and an additional 60 mm Hg upon voluntary contraction (maximum squeeze pressure) have been described to be necessary to maintain continence [3], single patients with markedly lower values reached above-average QoL scores. This result may be explained by the known low intra- and interindividual reproducibility of anorectal manometry values, particularly of maximum squeeze pressure, even under standard conditions in healthy volunteers [6]. An adaptation process may also contribute to good subjective QoL even with partially reduced sphincter function.

Of the frequent symptoms observed after radiotherapy for anal carcinoma, urgency of defecation was most closely associated with QoL, mirroring the clinical experience that the necessity of remaining close to a bathroom can severely limit patients' activities. Rectal bleeding was practically unrelated to QoL and may be a symptom more troublesome for the physician than the patient in this situation. Diarrhea was only weakly correlated with QoL in the present cohort.

Interestingly, assessment of fatigue in another single item of the GIQLI questionnaire revealed the highest correlation with QoL of all parameters analyzed. In recent years, the assessment of fatigue has reached tremendous importance in QoL research in cancer patients, including patients undergoing radiotherapy [12, 14, 15]. Whereas multidimensional fatigue questionnaires are available [11, 16], the possibility of measuring fatigue as a single item, such as scoring of worst fatigue during the last 24 h, has been explored [10]. Some authors consider fatigue the most important issue for cancer patients' QoL [23]. Whereas many physical and psychological factors have been described as causes of fatigue, one may speculate that in the present study the high correlation of QoL and fatigue may be a result of detrimental effects of impaired continence on both parameters.

Only limited data concerning the association of organ function and QoL in anal carcinoma is available from the literature. In the only previous investigation of anorectal manometry after radiotherapy for this disease, resting pressure and maximum squeeze pressure were also found to be reduced in the 8 patients investigated, 4 of them exhibiting clinically relevant partial incontinence [10]. QoL, however, was not assessed in this study.

The recently introduced colorectal-cancer-specific European Organization for Research and Treatment of Cancer (EORTC) QLQ-CR30 [19] was applied in an investigation of 41 anal carcinoma patients with a functional anal sphincter 10 years (median) after radiotherapy [1]. Whereas global QoL was similar to that of the general population, single items such as diarrhea, but not fatigue, were more severe in the anal carcinoma group. As in the present study, anal function (scored according to the Memorial Sloan-Kettering anal function criteria) was significantly associated with global QoL and the symptom scale 'defecation problems'.

The impact of organ function on QoL parameters after pelvic radiotherapy has been more extensively documented in other, more frequent diseases. In a study of 154 patients treated with definitive radiotherapy for carcinoma of the prostate, organ-specific morbidity such as urinary incontinence and bowel distress was correlated with global QoL only in the univariate analysis, whereas more general dimensions, including fatigue, remained a statistically independent predictive factor for QoL in the multivariate analysis [9]. Fecal incontinence was not included in this study, but has now been shown to be a frequent symptom in carcinoma of the prostate [22, 24].

In conclusion, anorectal function seems to be a major determinant of QoL after radiotherapy for anal carcinoma and other pelvic diseases. The discrepancy between high QoL scores and reduced anorectal manometry scores in anal carcinoma patients is most likely due to patients' adaptation to symptoms. Specific organ- or function-related questionnaires may be more closely correlated to QoL than anorectal manometry results.

### References


1   Allal AS, Sprangers MAG, Laurencet F, et al: Assessment of long-term quality of life in patients with anal carcinomas treated by radiotherapy with or without chemotherapy. Br J Cancer 1999; 80:1588–1594.
2   Bartelink H, Roelofsen F, Eschwege F, et al: Concomitant radiotherapy and chemotherapy is superior to radiotherapy alone in the treatment of locally advanced anal cancer: Results of a phase III randomized trial of the European Organization for Research and Treatment of Cancer Radiotherapy and Gastrointestinal Cooperative Groups. J Clin Oncol 1997;15:2040–2049.
3   Broens P, van Limbergen E, Penninckx F, et al: Clinical and manometric effects of combined external beam irradiation and brachytherapy for anal cancer. Int J Colorectal Dis 1998;13:68–72.

4 Eypasch E, Williams JI, Wood-Dauphinee S, et al: Gastrointestinal Quality of Life Index: Development, validation and application of a new instrument. Br J Surg 1995;82:216–222.

5 Flam M, John M, Pajak TF, et al: Role of mitomycin in combination with fluorouracil and radiotherapy, and of salvage chemoradiation in the nonsurgical treatment of epidermoid carcinoma of the anal canal: Results of a phase III randomized intergroup study. J Clin Oncol 1996;14: 2527–2539.

6 Freys SM, Fuchs KH, Fein M, et al: Inter- and intraindividual reproducibility of anorectal manometry. Langenbecks Arch Surg 1998;383:325–329.

7 Hu K, Minsky BD, Cohen AM, et al: 30 Gy may be an adequate dose in patients with anal cancer treated with excisional biopsy followed by combined-modality therapy. J Surg Oncol 1999;70: 71–77.

8 Kölbl O, Bratengeier K, Richter S, et al: Intracavitary afterloading therapy as a new technique of boost irradiation in anal canal carcinoma. Strahlenther Onkol 1997;173:513–518.

9 Lilleby W, Fossa SD, Waehre H, et al: Long-term morbidity and quality of life in patients with localized prostate cancer undergoing definitive radiotherapy or radical prostatectomy. Int J Radiat Oncol Biol Phys 1999;43:735–742.

10 Mendoza TR, Wang XS, Cleeland CS, et al: The rapid assessment of fatigue severity in cancer patients. Cancer 1999;85:1186–1196.

11 Piper BF, Lindsey AM, Dodd MJ, et al: The development of an instrument to measure the subjective dimension of fatigue; in Funk S, Tornquist E, Champagne M, et al (eds): Key Aspects of Comfort: Management of Pain, Fatigue and Nausea. New York, Springer, 1989, pp 199–208.

12 Richardson A: Fatigue in cancer patients: A review of the literature. Eur J Cancer Care 1995; 4:20–32.

13 Sailer M, Bussen D, Debus ES, et al: Quality of life in patients with benign anorectal disorders. Br J Surg 1998;85:1716–1719.

14 Smets EM, Visser MR, Willems-Groot AF, et al: Fatigue and radiotherapy A. Experience in patients undergoing treatment. Br J Cancer 1998;78:899–906.

15 Smets EM, Visser MR, Willems-Groot AF, et al: Fatigue and radiotherapy. B. Experience in patients 9 months following treatment. Br J Cancer 1998;78:907–912.

16 Smets EM, Garssen B, Bonke B, et al: The multidimensional fatigue inventory (MFI): Psychometric qualities of an instrument to assess fatigue. J Psychosom Res 1995;39:315–325.

17 Sprangers MAG, te Velde A, Aaronson NK: The construction and testing of the EORTC colorectal cancer-specific quality of life questionnaire module (QLQ-CR38). Eur J Cancer 1999;35:238–247.

18 Staib L, Gottwald T, Lehnert T, et al: Sphincter-saving treatment in epidermoid anal cancer: Cooperative analysis of 142 patients in five German university surgical centers. Int J Colorectal Dis 2000;15:282–290.

19 UICC: TNM Classification of Malignant Tumours. Berlin, Springer, 1997.

20 UKCCCR Anal Canal Cancer Trial Working Party: Epidermoid anal cancer: Results from the UKCCCR randomised trial of radiotherapy alone versus radiotherapy, 5-fluorouracil, and mitomycin. Lancet 1996;348:1049–1054.

21 Vordermark D, Sailer M, Flentje M, et al: Curative-intent radiation therapy in anal carcinoma: Quality of life and sphincter function. Radiother Oncol 1999;52:239–243.

22 Vordermark D: Is there more than one radiation-induced fecal incontinence syndrome? Re: Yeoh EK et al. (letter). Int J Radiat Oncol Biol Phys 2001;49:280–281.

23 Wolfe G: Fatigue: The most important consideration for the patient with cancer. Caring 2000; 19:46–47.

24 Yeoh EEK, Botten R, Russo A, et al: Chronic effects of therapeutic irradiation for localized prostatic carcinoma on anorectal function. Int J Radiat Oncol Biol Phys 2000;47:915–924.

Dirk Vordermark, MD, Department of Radiation Oncology, University of Würzburg,
Josef-Schneider-Strasse 11, D–97080 Würzburg (Germany)
Tel. +49 931 2015891, Fax +49 931 2012221,
E-Mail vordermark@strahlentherapie.uni-wuerzburg.de

Dörr W, Engenhart-Cabillic R, Zimmermann JS (eds): Normal Tissue Reactions in Radiotherapy and Oncology. Front Radiat Ther Oncol. Basel, Karger, 2002, vol 37, pp 140–150

# Normal Tissue Reactions after Linac-Based Radiosurgery and Stereotactic Radiotherapy

*M.W. Gross, R. Engenhart-Cabillic*

Klinik für Strahlentherapie und Radioonkologie, Uniklinik Marburg, Germany

Radiosurgery is a technique to deliver a single fraction of ionizing radiation to a precisely localized intracranial volume of pathological tissue [40]. The goal of radiosurgery is a very steep dose falloff at the treatment field margins. This results in a very well circumscribed high-dose region where radiation effects should be seen exclusively.

The steepness of this dose gradient depends on the size of the collimator. Dose falloff is more rapid for small beams and decreases linearly with increasing collimator diameter. The volume of normal tissue exposed to a high radiation dose increases even more steeply. Therefore, the size of radiosurgically treatable lesions is limited to a volume of 15 ml, corresponding to a diameter of 3 cm.

In contrast to conventionally fractionated radiotherapy the intrinsic radiosensitivity of the target plays only a minor role. In fact, the usual 'four Rs of radiotherapy' (recovery, repopulation, redistribution, reoxygenation) do not strictly apply to radiosurgery. Rather, the results of single, high radiosurgical doses are primarily a consequence of vascular effects [6] or antiproliferative effects. This leads to obliteration in arteriovenous malformations (AVMs) and to reproductive cell death in benign and malignant tumors, respectively.

The sparing effect of fractionation in the treatment of benign tumors in vivo is not as well categorized as for malignant tumors. Benign lesions mostly grow slowly and also respond slowly, if at all, to radiotherapy. Many studies have shown that fractionated radiotherapy as well as radiosurgery can prevent meningioma and neuroma regrowth for a long period of time [2, 23, 31, 39, 44, 53, 64, 65, 70].

This article describes the spectrum of neurotoxicity pertaining to the current variety of indications for which radiosurgery is being used. Additionally, a comparison with fractionated radiotherapy for benign tumors will be made as regards efficacy and toxicity of treatment.

## Radiobiological Considerations

Mature neurons are considered the most radioresistant cells in the CNS; glial and endothelial cells, representing the connective tissue supplying and stabilizing the neurons, are more radiosensitive. Damage of these cells will consecutively result in damage to the neural tissue. Already 1 day after high-dose irradiation a dose-dependent extravasation of serum proteins occurs as a sign of an impaired blood-brain barrier. This effect increases for some days, and is resolved during the subsequent weeks [48]. Two to 3 months after irradiation, reversible demyelinization represents the subacute reaction. Demyelinated plaques become confluent in a dose-dependent manner. Moreover, damage to capillaries occurs [5]. Subsequently neurons, glial cells and vessels display further alterations. Increasing gliosis and demyelination are associated with vascular changes [33]. Consequently, progressive vasogenic edema develops, which causes further impairment of cellular nutrition, and thus establishes a vicious circle.

To understand the clinical implications for the efficacy and toxicity of radiosurgery, the biological differences between early- and late-responding tissues must be considered. According to the linear-quadratic model a single dose of 20 Gy corresponds to a conventionally fractionated dose of about 50 Gy for early-responding and of 90–100 Gy for late-responding tissues (fig. 1). The target of radiosurgery may be early- or late-responding, while the surrounding normal structures (brain, cranial nerves, vasculature) are always late-responding. AVMs also appear to be late-responding, and radiation effects are highly dependent on dose per fraction.

Circumscript radionecroses or leukencephalopathies following conventionally fractionated irradiation are a typical late effect which can be seen after 9–36 months [16, 36, 48]. In contrast, after radiosurgery manifestation of necrosis is possible within only a few weeks [35].

The risk of development of this adverse reaction strongly correlates with the volume of irradiated brain and the applied dose [15, 17, 19]. Tolerance doses for some CNS structures are listed in table 1. Particularly the brain volume irradiated with more than 10 Gy is a predictive parameter [66, 68].

The sensitivity of the optic apparatus seems to be relatively high. Applying 10–22 Gy results in a complication rate of 20%. The tolerance dose for a single fraction for the optic system is considered to be 8 Gy [38, 62]. Preexisting injury of the visual system increases the sensitivity to radiosurgery with reduced tolerance.

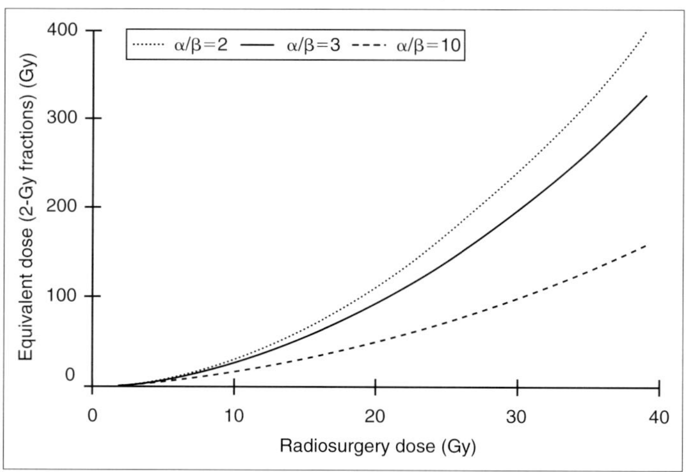

***Fig. 1.*** Fractionated dose (2-Gy fractions) versus radiosurgery dose of equal effectiveness for both early-responding ($\alpha/\beta = 10$) or late-responding tissue ($\alpha/\beta = 2$–3) according to the linear-quadratic model.

***Table 1.*** Tolerance doses of intracranial structures

| Structure | Tolerance dose Gy |
| --- | --- |
| Nerve II | 8 |
| Nerves III, IV, VI | 15–18 |
| Nerves V[a], VII[a] | 15 |
| Nerve VIII[a] | 12–15 |
| Optic chiasm | 8 |
| Optic tract | 8–12 |
| Sensoric and motoric cortex | 15–18 |
| Brainstem | 12 |

[a]Depending on irradiated nerve length.

## Symptoms and Time Course

The main acute reactions after radiosurgery, with a frequency of 0–7% [11, 42, 57], are headache, nausea and vomiting, and seizures. Usually these symptoms decline within a few hours. Probability and extent of these effects depend on lesion size and localization as well as duration of treatment. A headache can develop due to the fixation of the stereotactic headring at the skull. Symptoms will increase with duration of fixation. Irradiation of structures near the area postrema with doses >2.75 Gy in particular causes nausea and vomiting [42]. Antagonists of HT3 receptors may be administered prophylactically [4]. Seizures usually develop only in patients with a positive case history [11, 21].

Up to one third of radiosurgically treated patients display neuroradiological changes in MRI. Hyperintensity in T2-weighted sequences is most common. Pathological contrast enhancement is a sign of a disturbed blood-brain barrier. The frequency of these alterations in MRI is dependent on the kind of lesion treated. It amounts to 26–32% for AVM and to 6–20% after treatment of metastases and acoustic neuromas [11, 19, 34, 52]. These effects can be observed after 2–24 months. In about 30% they are symptomatic. In most cases a complete restitution of clinical symptoms occurs while the changes in imaging persist [42].

Differentiation between radionecrosis and recurrence of malignant tumors remains difficult. In CT or MRI pathological contrast enhancement and perifocal edema are very similar. In some cases the edema surrounding radionecrosis is less space-occupying. Even functional examinations like PET or SPECT only provide suggestions to solve this diagnostic problem. A comparison of imaging results with the clinical time course is essential.

Radionecrosis of normal brain tissue is the most important chronic side effect after irradiation. The clinical significance of a radiogenic necrosis depends on its localization. A high risk of morbidity and mortality results from a loss of function in important eloquent motoric or sensoric areas.


## Classification of Side Effects

Acute and chronic toxicity should be classified with respect to reversibility rather than to time course. According to the Radiation Therapy Oncology Group (RTOG) and EORTC, acute effects occur ≤90 days after completion of treatment, and chronic sequelae >90 days after onset of treatment. The grade definitions are summarized in table 2. Mild, moderate, severe (life-threatening) and fatal toxicity (death or complete loss of organ function) must be categorized [57, 58]. Grade 0 means absence of toxicity.

A classification of acute side effects after radiotherapy was established in Germany by Seegenschmiedt and Sauer [60] in 1993. This system is based on common toxicity criteria and encompasses all important organs and organ systems. Neurological function is described by the sensoric system, motoric system, consciousness, coordination, frame of mind, headache, behavior, and dizziness. Compatible with the EORTC/RTOG protocol, the common toxicity criteria score includes four grades of severity. Updates are given regularly.

Chronic adverse effects are characterized by the LENT/SOMA classification to record subjective and functional parameters [14]. However, this system

*Table 2.* Radiation Therapy Oncology Group (RTOG) CNS toxicity criteria

| Toxicity score | Description |
| --- | --- |
| 0 | no toxicity |
| 1 | mild neurological symptoms, no medication required |
| 2 | moderate neurological symptoms; outpatient medication required |
| 3 | severe neurological symptoms; outpatient or inpatient medication required |
| 4 | life-threatening neurological symptoms, including coma, paralysis, status epilepticus, radionecrosis requiring operation |
| 5 | death |

*Table 3.* Gardner-Robertson score of hearing function [25]

| Score | Description | Speech discrimination % | Pure tone hearing loss dB |
| --- | --- | --- | --- |
| I | good | 70–100 | 0–30 |
| II | useful | 50–69 | 31–50 |
| III | not useful | 5–49 | 51–90 |
| IV | minimal | 1–4 | 91 to maximum |
| V | deaf | 0 | not testable |

appears too inaccurate to assess subtle changes in several neurological functions. Acoustic or visual impairment requires even more sophisticated scoring systems. The functional integrity of the auditory organ not only depends on the displacement of the audibility threshold, but also on speech discrimination. Therefore, dedicated function tests were developed, such as the Gardner-Robertson score (see table 3) for hearing [25] and the House-Brackmann score for facial nerve function [27].

## Target-Specific Side Effects

### Arteriovenous Malformations

The management of AVM has been largely modified in the last decades due to the availability of endoscopic surgery techniques, endovascular embolization and radiosurgery. Besides preservation or improvement of neurological function the need for treatment of AVMs is based on the risk of bleeding. The aim of AVM treatment is the complete elimination of the nidus. Currently it is well established that the entire nidus has to be enclosed in the target volume and that

***Table 4.*** Obliteration rate and frequency of side effects in radiosurgical AVM treatment

| Reference No. | Number of patients | Obliteration % | Side effects % |
| --- | --- | --- | --- |
| 7 | >400 | 83 | 4.3 |
| 8 | 115 | 82 | 10 |
| 19 | 332 | 71 | 8.1 |
| 22 | 158 | 81 | 1.3 |
| 50 | 138 | 83 | n.a. |
| 55 | 66 | 86 | 6.7 |
| 69 | 30 | 71 | 0 |

n.a. = Not available.

no safety margin is required. Neither supporting arteries nor draining veins need to be irradiated [12]. The definition of the target volume remains one of the crucial steps in radiosurgical treatment of cerebral AVMs. To delineate the nidus, biplanar angiography is the gold standard in AVM treatment planning. CT angiography and MRI under stereotactic conditions are necessary to provide the spatial information. This leads to a more precise three-dimensional target localization. In most cases of AVMs the nidus is irregularly shaped. Therefore, either a spatial placement of several isocenters to cover the entire lesion with the prescribed isodose or micromultileaf collimators to adapt the beam shape to the lesion are required. For small and intermediate-sized AVMs the radiosurgical obliteration rate is about 80%. Retrospective analyses showed incomplete or wrong delineation of the nidus, and insufficient doses as main causes of failure [9, 24]. Further reasons for obliteration failure after radiosurgery are incomplete irradiation of large AVMs, type of angioarchitecture, recanalization of previously embolized regions, and others (25%).

In about 5% permanent clinically manifest side effects are seen even if the treated volume remains below 15 ml. In a study of Miyawaki et al. [46] the incidence of postradiosurgical MRI/T2 abnormalities (hyperintensity) was 72% and the incidence of radiation necrosis requiring resection was 22% when the nidus volume was 14–143 ml with a dose ≥16 Gy. Table 4 summarizes results from other studies. More frequently, temporary changes in patient status or in MRI signal characteristics are observed.

The risk of side effects and radionecrosis correlates well with the Spezler and Martin score [61]. A higher grade is associated with an increased risk. In a series of 120 patients, Engenhart-Cabillic and Debus [13] observed no clinical deterioration after radiosurgery of grade I and II AVMs but in 25% of grade V AVMs.

***Table 5.*** Frequency of side effects and tumor control for different fractionation schedules in the treatment of acoustic neuromas

| Reference No. | Number of patients | Fractionation | Side effects % | Response % |
|---|---|---|---|---|
| 23 | 12 | 32 × 2 Gy | 0 | 100 |
| 31 | 39 | 22 × 2 Gy + 4 | 16 | 97 |
| 32 | 19 | 5–6 × 6 Gy | 0 | 100 |
| 39 | 38 | 5(4) × 4(5) Gy | 0 | 100 |
| 44 | 37 | 5 × 4–5 Gy | 3 | 91 |
| 53 | 33 | 3 × 7 Gy | 9 | 97 |
| 65 | 12 | 30 × 1.8 Gy | 8 | 100 |

*Acoustic Neurinomas*

Major parameters of the functional outcome after radiosurgery of acoustic neurinoma are tumor size and tumor site. The length of the irradiated cranial nerves as well as the minimum dose are prognostic factors [41, 45]. The trigeminal nerve is only altered by the extracanalicular part of the neuroma, the vestibulocochlear nerve and the facial nerve by the extra- and intracanalicular part. Radiosensitivity increases from the facial and trigeminal to the vestibulocochlear nerve. Hearing loss is not correlated with loss of vestibular function [30]. Flickinger et al. [18] developed a formula to calculate the risk of neuropathy by dose and length irradiated.

Neuroradiological changes in MRI can usually only be observed if the brainstem is affected, which can be seen as hyperintensity in $T_2$-weighted sequences in 6–20% of patients [19, 34]. A disturbance of the blood-brain barrier with contrast enhancement is less frequent. Only in a quarter of these symptomatic cases does clinical deterioration occur; temporary steroid medication usually improves or resolves the symptoms.

Fractionated stereotactic irradiation especially for large tumors was established by several authors using different fractionation schemes [3, 23, 31, 32, 39, 44, 53, 65]. Dose per fraction ranged from 2 to 7 Gy with total doses between 31 and 64 Gy. Side effects concerning hearing and cranial nerve function were seen in about 6% ranging from 0 to 16%. Details are shown in table 5. In comparison with radiosurgery and neurosurgery, fractionated stereotactic radiotherapy seems to be more tissue-sparing (table 6).

*Metastases*

Acute reactions, like nausea and/or vomiting or seizures, following radiosurgery of metastases are observed in 4–9%. However, epileptic convulsions usually occur only in patients with a positive case history [11, 21]. Therefore, an increase of anticonvulsive medication is sufficient.

***Table 6.*** Results of radiosurgery, neurosurgery and fractionated radiotherapy (RT)

|  | Radiosurgery % | Neurosurgery % | Fractionated RT % |
| --- | --- | --- | --- |
| Response | 85–98 | 92–98 | 88–100 |
| Recurrence rate | 2–10 | ~7 | 0–12 |
| Unchanged hearing | 30–70 | 14–57 | 50–100 |
| Useful hearing | –85 | 36–78 | 66–100 |
| Toxicity to nerves V, VII | 0–20 | ~8 | 0–16 |

Minor side effects such as transient edema are observed in 18% of patients after 2–4 months. Usually the resulting symptoms are effectively treated by a temporary administration of steroids [42, 52].

Severe chronic symptomatically adverse reactions occur in 0–8% of patients [1, 35, 42, 52, 59]. In most cases steroid medication leads to complete recovery from the symptoms. Pirzkall et al. [52] reported 236 patients bearing 311 metastases. Only in 4 patients (1.3%) symptomatic radionecrosis was observed. In one case reoperation was necessary, after which radionecrosis was confirmed histologically.

*Meningiomas and Pituitary Adenomas*

Meningiomas show excellent local control after radiosurgery [10, 26, 37, 56, 64] as well as after fractionated stereotactic radiotherapy [2, 43, 49]. Patients with larger tumors close to critical structures, e.g. brainstem or cranial nerves, often develop side effects, with a rate ≤42% [10, 26, 63]. The damage of the cranial nerves usually develops within 3–31 months, and is reversible in a substantial proportion of patients. Careful planning and restriction of the lesion size lead to a reduction of radiation-induced adverse effects to about 5% [28, 37, 56, 64]. Therefore, in selected cases, radiosurgery is a reasonable treatment method for benign tumors, particularly of the skull base.

The potential toxicity of radiosurgery of pituitary adenomas on one hand is damage to the cranial nerves II–VI and, on the other hand, pituitary insufficiency [54, 67]. The treatment of choice for microadenomas is microsurgical resection. This procedure results in rapid normalization of elevated hormone levels. In contrast, the latency period for an endocrine response following radiosurgery is in the range of 1–2 years [29, 51], and hence is slightly shorter than after conventionally fractionated radiotherapy [47, 70]. Comparison of radiosurgery and stereotactic radiotherapy by Yoon et al. [70] revealed 27% adverse effects of neural function and 23% pituitary insufficiency after radiosurgery versus 0 and 20% after stereotactic radiotherapy, respectively. Therefore, the lower morbidity after fractionated stereotactically guided radiotherapy seems to be preferable.

## References

1 Alexander E 3rd, Moriarty TM, Davis RB, et al: Stereotactic radiosurgery for the definitive, noninvasive treatment of brain metastases. J Natl Cancer Inst 1995;87:34–40.

2 Alheit H, Saran FH, Warrington AP, et al: Stereotactically guided conformal radiotherapy for meningiomas. Radiother Oncol 1999;50:145–150.

3 Becker G, Schlegel W, Major J, Grote EH, Bamberg M: Stereotactic convergent beam radiosurgery versus stereotactic conformation beam radiotherapy. Acta Neurochir Suppl (Wien) 1995;63:44–51.

4 Bodis S, Alexander E 3rd, Kooy H, et al: The prevention of radiosurgery-induced nausea and vomiting by ondansetron: Evidence of a direct effect on the central nervous system chemoreceptor trigger zone. Surg Neurol 1994;42:249–252.

5 Chang SD, Adler JR: Treatment of cranial base meninigiomas with linear accelerator radiosurgery. Neurosurgery 1997;41:1019–1027.

6 Chang SD, Shuster DL, Steinberg GK, et al: Stereotactic radiosurgery of arteriovenous malformations: Pathologic changes in resected tissue. Clin Neuropathol 1997;16:111–116.

7 Debus J, Pirzkall A, Schlegel W, Wannenmacher M: Stereotaktische Einzeitbestrahlung (Radiochirurgie) – Methodik, Indikationen, Ergebnisse. Strahlenther Onkol 1999;175:47–56.

8 Deruty R, Pelissou-Guyotat I, Morel C, et al: Reflections on the management of cerebral arteriovenous malformations. Surg Neurol 1998;50:245–255.

9 Ellis TL, Friedman WA, Bova FJ, et al: Analysis of treatment failure after radiosurgery for arteriovenous malformations. J Neurosurg 1998;89:104–110.

10 Engenhart R, Kimmig BN, Hover KH, et al: Stereotactic single high dose radation therapy of benign intracranial meningiomas. Int J Radiat Oncol Biol Phys 1990;19:1021–1026.

11 Engenhart R, Kimmig BN, Hover KH, et al: Long-term follow-up for brain metastases treated by percutaneous stereotactic single high-dose irradiation. Cancer 1993;71:1353–1361.

12 Engenhart R, Wowra B, Debus J, et al: The role of high-dose, single-fraction irradiation in small and large intracranial arteriovenous malformations. Int J Radiat Oncol Biol Phys 1994;30:521–529.

13 Engenhart-Cabillic R, Debus J: Linear accelerator radiosurgey for arteriovenous malformations: The relationship of size, dose, time, and planning factors onto outcome; in Hellwig D, Bauer BL (eds): Minimally Invasive Techniques for Neurosurgery. Berlin, Springer, 1998, pp 145–150.

14 EORTC/RTOG Late Effects Working Group: LENT SOMA tables. Radiother Oncol 1995;35: 17–60.

15 Fabrikant JI, Lyman JT, Hosobuchi Y: Stereotactic heavy-ion Bragg peak radiosurgery for intracranial vascular disorders: Method for treatment of deep arteriovenous malformations. Br J Radiol 1984;57:479–490.

16 Fajardo LF: Morphology of radiation effects on normal tissue; in Perez CA, Brady LW (eds): Principles and Practice of Radiation Oncology, ed 2. Philadelphia, Lippincott, 1992, pp 114–123.

17 Flickinger JC, Schell MC, Larson DA: Estimation of complications for linear accelerator radiosurgery with the integrated logistic formula. Int J Radiat Oncol Biol Phys 1990;19:143–148.

18 Flickinger JC, Kondziolka D, Lunsford LD: Dose and diameter relationships for facial, trigeminal, and acoustic neuropathies following acoustic neuroma radiosurgery. Radiother Oncol 1996;41:215–219.

19 Flickinger JC, Kondziolka D, Maitz AH, et al: Analysis of neurological sequelae from radiosurgery of arteriovenous malformations: How location affects outcome. Int J Radiat Oncol Biol Phys 1998; 40:273–278.

20 Friedman WA, Bova FJ: Linear accelerator radiosurgery for arteriovenous malformations. J Neurosurg 1992;77:832–841.

21 Friedman WA: Radiosurgery for arteriovenous malformations. Clin Neurosurg 1995;42:328–347.

22 Friedman WA, Blatt DL, Bova FJ, Buatti JM, Mendenhall WM, Kubilis PS: The risk of hemorrhage after radiosurgery for arteriovenous malformations. J Neurosurg 1996;84:912–919.

23 Gademann G, Schlegel W, Debus J, et al: Fractionated stereotactically guided radiotherapy of head and neck tumors: A report on clinical use of a new system in 195 cases. Radiother Oncol 1993;29: 205–213.

24 Gallina P, Merienne L, Meder JF, et al: Failure in radiosurgery treatment of cerebral arteriovenous malformations. Neurosurgery 1998;42:996–1002.

25  Gardner G, Robertson JH: Hearing preservation in unilateral acoustic tumor surgery. Ann Otol Rhinol Laryngol 1988;97:55–66.

26  Hakim R, Alexander E 3rd, Loeffler J, et al: Results of linear accelerator-based radiosurgery for intracranial meningiomas. Neurosurgery 1998;42:446–453.

27  House JW, Brackmann DE: Facial nerve grading system. Otolaryngol Head Neck Surg 1985;93: 146–147.

28  Hudgins WR, Barker JL, Schwartz DE, Nichols TD: Gamma knife treatment of 100 consecutive meningiomas. Stereotact Funct Neurosurg 1996;66(suppl):121–128.

29  Ikeda H, Jokura H, Yoshimoto T: Gamma knife radiosurgery for pituitary adenomas: Usefulness of combined transsphenoidal and gamma knife radiosurgery for adenomas invading the cavernous sinus. Radiat Oncol Investig 1998;6:26–34.

30  Ito K, Kurita H, Sugasawa K, et al: Neuro-otological findings after radiosurgery for acoustic neurinomas. Arch Otolaryngol Head Neck Surg 1996;122:1229–1233.

31  Kagei K, Shirato H, Suzuki K, et al: Small-field fractionated radiotherapy with or without stereotactic boost for vestibular schwannoma. Radiother Oncol 1999;50:341–347.

32  Kalapurakal JA, Silverman CL, Akhtar N, et al: Improved trigeminal and facial nerve tolerance following fractionated stereotactic radiotherapy for large acoustic neuromas. Br J Radiol 1999;72: 1202–1207.

33  Karger CP, Hartmann GH, Peschke P, et al: Dose-response relationship for late functional changes in the rat brain after radiosurgery evaluated by magnetic resonance imaging. Int J Radiat Oncol Biol Phys 1997;39:1163–1172.

34  Kobayashi T, Tanaka T, Kida Y: The early effects of gamma knife on 40 cases of acoustic neurinoma. Acta Neurochir Suppl (Wien) 1994;62:93–97.

35  Kocher M, Voges J, Müller RP, et al: Linac radiosurgery for patients with a limited number of brain metastases. J Radiosurg 1998;1:9–15.

36  van der Kogel AJ: Central nervous system radiation injury in small animal models; in Gutin PH, Leibel SA, Sheline GE (eds): Radiation Injury to the Nervous System. New York, Raven Press, 1991, pp 91–112.

37  Kondziolka D, Levy EI, Niranjan A, et al: Long-term outcomes after meningioma radiosurgery: Physician and patient perspectives. J Neurosurg 1999;91:44–50.

38  Leber KA, Berglöff J, Langmann G, et al: Radiation sensitivity of visual and oculomotor pathways. Stereotact Funct Neurosurg Suppl 1995;64:233–238.

39  Lederman G, Lowry J, Wertheim S, et al: Acoustic neuroma: Potential benefits of fractionated stereotactic radiosurgery. Stereotact Funct Neurosurg. 1997;69:175–182.

40  Leksell L: The stereotactic method and radiosurgery of the brain. Acta Chir Scand 1951;102:316–319.

41  Linskey ME, Flickinger JC, Lunsford LD: Cranial nerve length predicts the risk of delayed facial and trigeminal neuropathies after acoustic tumor stereotactic radiosurgery. Int J Radiat Oncol Biol Phys 1993;25:227–233.

42  Loeffler JS, Alexander E 3rd: Radiosurgery for the treatment of intracranial metastases; in Alexander E 3rd, Loeffler JS, Lunsford LD (eds): Stereotactic Radiosurgery. New York, McGraw-Hill, 1993, pp 197–206.

43  Maguire PD, Clough R, Friedman AH, Halperin EC: Fractionated external-beam radiation therapy for meningiomas of the cavernous sinus. Int J Radiat Oncol Biol Phys 1999;44:75–79.

44  Meijer OW, Wolbers JG, Baayen JC, Slotman BJ: Fractionated stereotactic radiation therapy and single high-dose radiosurgery for acoustic neuroma: Early results of a prospective clinical study. Int J Radiat Oncol Biol Phys 2000;46:45–49.

45  Miller RC, Foote RL, Coffey RJ, et al: Decrease in cranial nerve complications after radiosurgery for acoustic neuromas: A prospective study of dose and volume. Int J Radiat Oncol Biol Phys 1999;43:305–311.

46  Miyawaki L, Dowd C, Wara W, et al: Five year results of LINAC radiosurgery for arteriovenous malformations: Outcome for large AVMS. Int J Radiat Oncol Biol Phys 1999;44:1089–1106.

47  Mitsumori M, Shrieve DC, Alexander E 3rd, et al: Initial clinical results of LINAC-based stereotactic radiosurgery and stereotactic radiotherapy for pituitary adenomas. Int J Radiat Oncol Biol Phys 1998;42:573–580.

48    Nakata H, Yoshimine T, Murasawa A, et al: Early blood-brain barrier disruption after high-dose single-fraction irradiation in rats. Acta Neurochir (Wien) 1995;136:82–86.

49    Nutting C, Brada M, Brazil L, et al: Radiotherapy in the treatment of benign meningioma of the skull base. J Neurosurg 1999;90:823–827.

50    Oppenheim C, Meder JF, Trystram D, et al: Radiosurgery of cerebral arteriovenous malformations: Is an early angiogram needed? Am J Neuroradiol 1999;20:475–481.

51    Park YG, Chang JW, Kim EY, Chung SS: Gamma knife surgery in pituitary microadenomas. Yonsei Med J 1996;37:165–173.

52    Pirzkall A, Debus J, Lohr F, et al: Radiosurgery alone or in combination with whole brain radiotherapy for brain metastases. J Clin Oncol 1998;16:3563–3569.

53    Poen JC, Golby AJ, Forster KM, et al: Fractionated stereotactic radiosurgery and preservation of hearing in patients with vestibular schwannoma: A preliminary report. Neurosurgery 1999;45: 1299–1305.

54    Pollock BE, Kondiolka D, Lunsford LD, Flickinger JC: Stereotactic radiosurgery for pituitary adenomas: Imaging, visual and endocrine results. Acta Neurochir Suppl (Wien) 1994;62:33–38.

55    Sasaki T, Kurita H, Saito I, et al: Arteriovenous malformations in the basal ganglia and thalamus: Management and results in 101 cases. J Neurosurg 1998;88:285–292.

56    Shafron DH, Friedman WA, Buatti JM, et al: Linac radiosurgery for benign meningiomas. Int J Radiat Oncol Biol Phys 1999;43:321–327.

57    Shaw EG, Coffey RJ, Dinapoli RP: Neurotoxicity of radiosurgery. Semin Radiat Oncol 1995;5: 235–245.

58    Shaw EG, Kline R, Gillin M, et al: Radiation Therapy Oncology Group: Radiosurgery quality assurance guidelines. Int J Radiat Oncol Biol Phys 1993;27:1231–1239.

59    Shaw E, Scott C, Souhami L, et al: Radiosurgery for the treatment of previously irradiated recurrent primary brain tumors and brain metastases: Initial report of radiation therapy oncology group protocol (90-05). Int J Radiat Oncol Biol Phys 1996;34:647–654.

60    Seegenschmiedt MH, Sauer R: Systematik der akuten und chronischen Strahlenfolgen. Strahlenther Onkol 1993;83:83–95.

61    Spezler RF, Martin NA: A proposed grading system for arteriovenous malformations. J Neurosurg 1986;65:476–483.

62    Tanaka T, Kobayashi T, Kida Y, et al: Comparison between adult and pediatric arteriovenous malformations treated by gamma knife radiosurgery. Stereotact Funct Neurosurg 1996;66(suppl 1): 288–295.

63    Tishler RB, Loeffler JS, Lunsford LD, et al: Tolerance of cranial nerves of the cavernous sinus to radiosurgery. Int J Radiat Oncol Biol Phys 1993;27:215–221.

64    Valentino V, Schinaia G, Raimondi A: The results of radiosurgical management of 72 middle fossa meningiomas. Acta Neurochir (Wien) 1993;122:60–70.

65    Varlotto JM, Shrieve DC, Alexander E 3rd, et al: Fractionated stereotactic radiotherapy for the treatment of acoustic neuromas: Preliminary results. Int J Radiat Oncol Biol Phys 1996;36:141–145.

66    Voges J, Treuer H, Sturm V, et al: Risk analysis of linear accelerator radiosurgery. Int J Radiat Oncol Biol Phys 1996;36:1055–1063.

67    Voges J, Sturm V, Deuss U, et al: LINAC-radiosurgery (LINAC-RS) in pituitary adenomas: Preliminary results. Acta Neurochir Suppl (Wien) 1996;65:41–43.

68    Voges J, Treuer H, Lehrke R, et al: Risk analysis of LINAC radiosurgery in patients with arteriovenous malformation (AVM). Acta Neurochir Suppl (Wien) 1997;68:118–123.

69    Wolbers JG, Mol HC, Kralendonk JH, et al: Stereotactic radiosurgery with adjusted linear accelerator for cerebral arteriovenous malformations: Preliminary results in the Netherlands. Ned Tijdschr Geneeskd 1999;143:1215–1221.

70    Yoon SC, Suh TS, Jang HS, et al: Clinical results of 24 pituitary macroadenomas with linac-based stereotactic radiosurgery. Int J Radiat Oncol Biol Phys 1998;41:849–853.

M.W. Gross, Klinik für Strahlentherapie und Radioonkologie, Uniklinik Marburg,
Baldingerstrasse, D–35033 Marburg, (Germany)
Tel. +49 6421 28 65853, Fax +49 6421 28 66426, E-Mail gross@mailer.uni-marburg.de

Dörr W, Engenhart-Cabillic R, Zimmermann JS (eds): Normal Tissue Reactions in Radiotherapy and Oncology. Front Radiat Ther Oncol. Basel, Karger, 2002, vol 37, pp 151–162

# Fitting of Tissue Tolerance Data to Analytic Function: Improving the Therapeutic Ratio

*Chandra M. Burman*

Memorial Sloan-Kettering Cancer Center, New York, N.Y., USA

There have been rapid developments in the planning and delivery of radiation therapy. There are three major areas where there have been significant improvements. (1) The use of multimodality imaging such as the computed tomography (CT), magnetic resonance imaging, positron emission tomography and ultrasound in localizing the diseased and normal tissues. (2) The development of CT-based 3-dimensional (3-D) treatment planning systems, where the target volume can be visualized from a beam's eye view perspective [21], allowing for significant reduction in dose to the normal tissues. With a modern 3-D treatment planning system it is possible to plan and calculate dose distributions with noncoplanar beams in three dimensions. The dose calculations are more precise because the effect of tissue inhomogeneity is incorporated in the dose distribution, which is particularly significant for treatment sites like lung. The output of a 3-D planning system is also more quantitative. Other than the planar isodose distributions, it provides dose volume information in the form of dose volume histograms (DVHs) [5]. (3) The treatment delivery is more efficient. Modern treatment accelerators are capable of delivering intensity-modulated beams [2], and the online imaging devices can easily verify the positional accuracy of the delivered dose [18].

In the past radiation oncologists have relied on the planar isodose distributions to evaluate the merit of a plan and to make the treatment decisions. However, as the treatment plans become more complex with unusual dose distributions, DVHs are useful in making the decisions. As the DVH information has become easily available there have been attempts to incorporate the DVH into the treatment decision process. Indices such as equivalent uniform dose [19], tumor control probability (TCP) [7, 20] and normal tissue complication probability (NTCP) [12] are attempts to condense and simplify the information. These can be used to make treatment decisions, compare competing plans, and optimize treatment plans. In this presentation the focus is on the determination of NTCP and how it can be used to improve the therapeutic ratio.

## Method of NTCP Calculation

There have been attempts to calculate NTCP based on models that rely on the functional architecture of the tissues: for example, the Schultheiss model [24] for tissues with serial architecture such as the cord and brain stem, and analysis of liver complication data by Jackson et al. [8] using a parallel architecture model. However, here we will focus on the use of a parametric model by Lyman [12].

*Lyman Model*
This phenomenological model with four parameters describes the tissue response for uniform partial organ irradiation. The NTCP is described as a function of dose ($D$) and partial volume as a set of equations:

$$\text{NTCP} = \frac{1}{\sqrt{2\pi}} \cdot \int_{-\infty}^{t} \exp(-x^2/2)\, dx \tag{1}$$

$$v = V/V_{\text{ref}} \tag{2}$$

$$t = (D - \text{TD}_{50}(v)) / (m \cdot \text{TD}_{50}(v)) \tag{3}$$

$$\text{TD}_{50}(v) = \text{TD}_{50}(1) \cdot v^{-n} \tag{4}$$

The above set of equations has four parameters: $V_{\text{ref}}$, $\text{TD}_{50}(1)$, $n$ and $m$. $V_{\text{ref}}$ is the reference volume; $\text{TD}_{50}(1)$ is the dose to the whole organ or the reference volume which will lead to complication probability of 50%; the parameter n determines the volume dependence of the complication probability; the slope of the NTCP as a function of dose is governed by the value of m. Figure 1 shows a set of curves for liver [4] with $\text{TD}_{50}(1) = 40\,\text{Gy}$, $n = 0.32$ and $m = 0.15$ for $\frac{1}{3}$, $\frac{2}{3}$ and whole organ irradiation. The NTCP can also be displayed as a surface when complication is graphed as a function of dose and partial volume, as presented in figure 2. This model requires the knowledge of reference volume, $\text{TD}_{50}(1)$, n and m for a given complication endpoint to determine the complication probability for uniform partial organ irradiation.

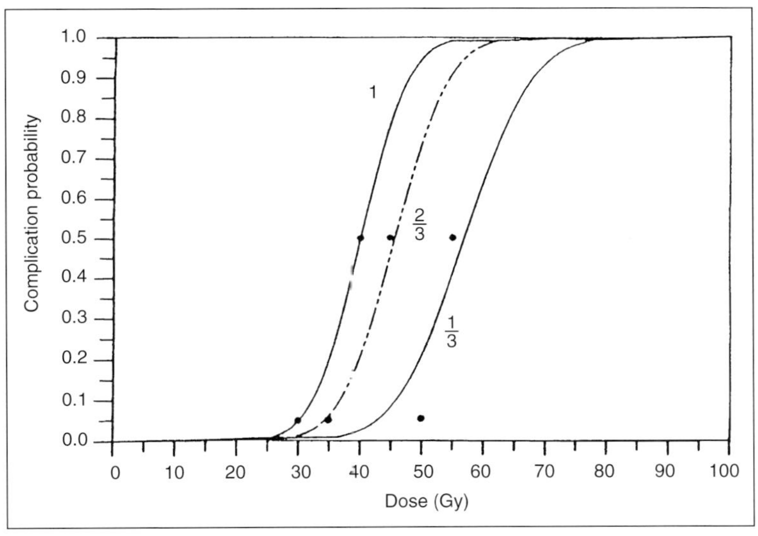

*Fig. 1.* NTCP as a function of dose for the liver for uniform irradiation of the whole, $\frac{2}{3}$ and $\frac{1}{3}$ organ.

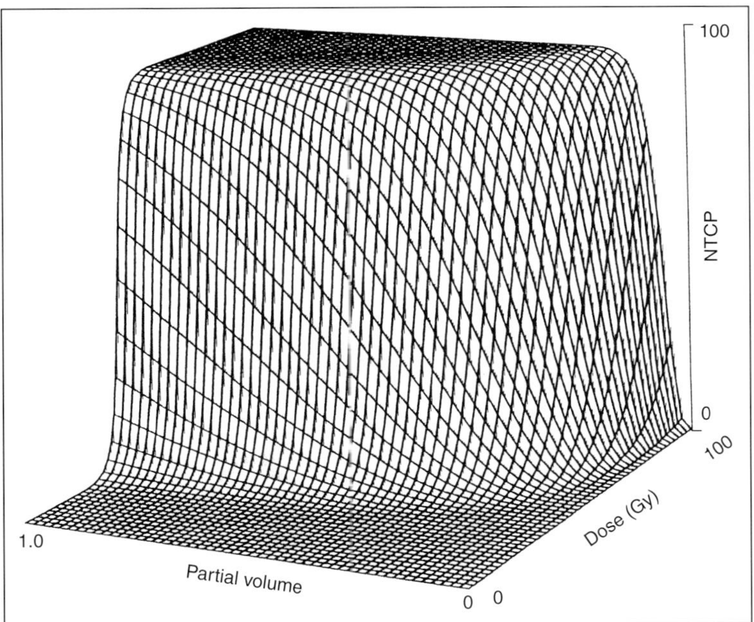

*Fig. 2.* NTCP as a function of dose and partial volume for uniform irradiation of partial organ.

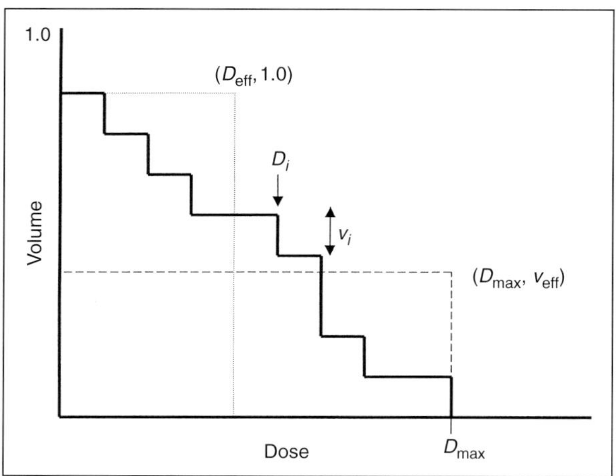


*Fig. 3.* Schematic diagram showing the conversion of a DVH for inhomogeneous dose distribution to equivalent DVH for uniform irradiation using the effective dose method ($D_{eff}$, 1) and the effective volume method ($D_{max}$, $v_{eff}$).

### Histogram Reduction Schemes

The Lyman model predicts the NTCP for partial volume uniform irradiation only; it cannot calculate the complication for an inhomogeneous dose distribution. There are histogram reduction schemes that can convert the DVH for an inhomogeneous distribution to an equivalent DVH for partial organ uniform irradiation. The scheme by Lyman and Wolbarst [13, 14] converts the cumulative DVH to an equivalent uniform effective dose, $D_{eff}$, to the reference volume such that the NTCP for the two distributions are equivalent. Whereas the effective volume method of Kutcher and Burman [10] reduces the DVH for nonuniform irradiation to a one-step histogram with a uniform dose, $D_{max}$, to the effective volume:

$$v_{eff} = \sum_i v_i (D_i/D_{max})^{1/n} \tag{5}$$

where $v_i$ is subvolume irradiated to dose $D_i$. Figure 3 shows an example of two types of reduction schemes.

### Determination of the Parameters

For a given endpoint, use of the Lyman model to calculate the NTCP requires the knowledge of $V_{ref}$, $TD_{50}(1)$, $n$ and $m$. The clinical data are needed to determine these parameters. For selected endpoints, an attempt was made by the Collaborative Working Group on the Evaluation of the Treatment Planning for External Beam Radiotherapy [22]. A task group headed by Emami et al. [6] was formed to search the literature and draw from their own clinical experience to provide the most up-to-date tolerance data for the selected tissues, with emphasis on the partial volume effects. The group reviewed the protocols for the eight treatment sites – nasopharynx, larynx, breast, lung, para-aortic nodes, prostate, and rectum – and identified a set of most serious dose-limiting end-points. These complications are generally taken into consideration during radio-therapy. Only the conventional schedule of 180–200 cGy per fraction at five treatments per week was considered. The reference volume was the whole organ for most of the tissues except for the spinal cord and skin with a reference length of 20 cm and reference area of 100 cm$^2$, respectively. The task group arbitrarily divided the reference volume of each organ into three categories: $\frac{1}{3}$, $\frac{2}{3}$ and the whole organ. The intention of the group was to assign appropriate tolerance doses to each of these volumes. Only adult tissue tolerance was considered. The task group reviewed the existing literature to collect the tolerance data; one of the major problems was the lack of quantitative dose-volume information. Most of the information was pre-CT and pre-3-D based on the simple field arrangements and on 2-dimensional isodose distributions. The group was able to come up with all 6 data points for only 11 endpoints, 4–5 data points for eight tissues, 2–3 points for 10 endpoints. For 8 tissues no volume dependence was given.

*Curve Fitting Procedure*

Due to the lack of a sufficient number of data points the decision was made to fit the curves 'by eye' rather than by a statistical method. The cases for which all 6 data points were available, a set of three curves (NTCP vs. dose for 3 partial volumes: 1, $\frac{2}{3}$ and $\frac{1}{3}$) was generated for given value of $TD_{50}(1)$ and estimated values of parameters n and m. If the fit was poor, first n was adjusted to obtain the best fit for the volume dependence. The next step was to vary the parameter m to make the probability curve pass through $TD_5(1)$ and $TD_{50}(1)$. While adjusting the parameters more weight was given to the data points for a 5% complication rate.

For the spinal cord, kidney, brain stem, bladder, lung, small intestine and colon tolerance doses for the whole organ irradiation – $TD_5(1)$ and $TD_{50}(1)$ – were provided along with at least one data point for partial volume irradiation. In these cases first, the parameter m was adjusted to obtain a good fit through the $TD_5(1)$ and $TD_{50}(1)$. Then the volume dependence parameter n was adjusted to obtain a good fit through the available partial volume data.

Only whole organ irradiation data was provided for the rectum, cauda equina, lens, retina, femoral head and neck, optical chiasm, optic nerve, brachial plexus and thyroid. For the rib cage the data for the $\frac{1}{3}$ of the volume was given. With these data it was possible to determine the values of $TD_{50}$ and *m*. To determine the value of the volume dependence parameter, $n$, for the organs with insufficient data, a best clinical estimate was made by a group of oncologists.

For the ear (middle/external) tolerance data for 2 endpoints, acute serous otitis and chronic serous otitis were provided. For both of these endpoints the tolerance data for the whole, $\frac{2}{3}$ and $\frac{1}{3}$ organ were the same which would lead to $n = 0$ (no volume dependence). This implies that a very small volume irradiated has the same complications as if the whole organ was irradiated, which is believed to be physically not correct. Therefore in these cases a very small value of 0.01 was assigned for the volume dependence parameter. A complete list of all four parameters is given in table 1 of reference 4.

## Refinement of the Parameters

With the help of the framework described above the NTCPs can be calculated from the DVHs. However, the accuracy or the reasonableness of the estimates depends on the clinical data used to determine the parameters. The cases where a substantial number of data points were provided for a wide range of partial volumes based on the clinical data lead to some confidence in the calculated values. For the cases where limited clinical data were available, estimates had to be made based on the oncologists' clinical experience; there is much less confidence in the calculated values. The objective was to develop a framework in which the NTCPs can be used for comparing competing plans and to refine planning techniques. It was anticipated that as the quantitative dose-volume data from 3-D planning system became widely available based on the new data the parameter could be refined to improve the accuracy of the predictions. The following are some of the recent attempts by the Michigan group.

### Liver

Lawrence et al. [11] analyzed radiation complication data for a group of 79 patients. DVHs were used to calculate the complication probability. Their results indicate that the parameters $TD_{50}(1) = 40\,Gy$ and $m = 0.15$ are the same as the earlier estimates [4]. However, the calculated value of $n$ was $0.69 \pm 0.06$ compared to 0.32 [4].

### Lung

3-D plans for 63 patients treated with radiation were analyzed by Martel et al. [16]. The group had 21 patients with Hodgkin's disease and 42 patients with non-small-cell lung cancer. The calculated complication probability was compared with the observed rate of clinical complication. Their results indicate that the original parameters [4] correlate with the actual complication rates for the Hodgkin's patients, but not as well for the lung cancer patients treated to larger volumes of lung and with high doses.

*Visual Pathway Structures*

A group of 20 patients with advanced paranasal sinus malignant tumors were evaluated with the 3-D treatment planning system and DVH analysis was done for chiasm, optic nerve and retina. Martel et al. [15] compared the observed rate of complication with the calculated risk. Their results find published parameters to be adequate for chiasm and retina. For the optic nerve $TD_{50}(1) = 72$ Gy provides a better match with the observed complication rate compared to $TD_{50}(1) = 65$ Gy [4].

*Future Developments*

It is anticipated that as the use of 3-D treatment planning systems increases, more clinical data will be collected that can be used for verification and modification of the existing parameters. In addition it is anticipated that clinical data will be collected for less severe endpoints. Although these complications, for example, grade 2 complication of rectal bleeding during prostate treatment, are not life-threatening, radiation oncologists still attempt minimization. Good quality clinical data with wide dose variation for different partial volumes are required for better modeling and understanding of the organ response to dose variation. As suggested by Moiseenko et al. [17], accumulation of human data on partial volume irradiation in an easily accessible data bank would be useful.

## Use of Biological Indices

As stated earlier, the biological indices TCP and NTCP can be used instead of dose to make the treatment decision. Some of the examples follow below.

*Plan Comparison*

With the increasing use of 3-D treatment planning systems, the treatment plans have become complex and sometimes the beam arrangements are unusual. Often the physician is presented with a set of competing plans to select from. An example is shown in figure 4. The cumulative DVHs for the rectal wall are displayed for two types of prostate plans. With the intensity-modulated radiotherapy (IMRT) plan more dose above 80 Gy is delivered to the tissues, whereas with the 3-D plan more tissue is irradiated at doses <80 Gy. For an experienced physician the answer may be obvious but to carry out a computer-based decision the calculation of NTCP is helpful. Based on the model

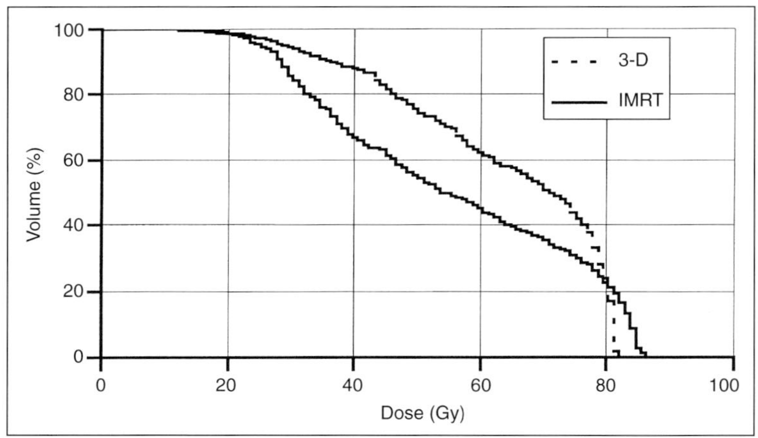


*Fig. 4.* Rectal wall DVHs for two types of plans for a prostate patient. With the IMRT plan more dose >80 Gy is delivered, whereas with the 3-D plan more tissue is irradiated at <80 Gy.

discussed above, for the rectum [$TD_{50}(1) = 80$ Gy, $n = 0.12$, $m = 0.15$] the NTCP for the IMRT plan is lower (5%) compared to the 3-D plan (6%).

*Guidance in Dose Escalation*

The use of 3-D planning for localized prostate cancer has reduced the rate of radiation-induced complications. There are dose escalation trials under way at many institutions to increase the control rate [26]. An example of the use of the NTCP and TCP to guide the dose escalation was presented by Burman et al. [3]. Two types of plans were performed for a group of 10 patients, for prescription doses from 75.6 to 95 Gy. Type I plan involved 6 fields (2 lateral, 2 anterior oblique and 2 posterior oblique), with the dose prescribed to maximum isodose line encompassing the planning target volume (PTV). Type II plan comprised a primary treatment of 72 Gy to the PTV (using the same 6 fields as for type I plan), and a boost with posterior oblique beams to deliver the additional dose, except to the anterior portion of the rectal wall. The TCP was calculated using Goitein's model [7] and the rectal wall NTCP were calculated based on the model described above. The probability of uncomplicated control [9, 23], defined as TCP × (1–NTCP), was also calculated. The results are shown in figure 5. For the type I plan the average TCP increases from 75% at 75.6 Gy to 98% at 95 Gy. The average rectal NTCP also increases to an unacceptably high level of >20% at 95 Gy. The probability of uncomplicated control initially increases, is maximum at 85 Gy, and then decreases for higher doses. The TCP for the type II plan is slightly reduced relative to that of type I plan. However,

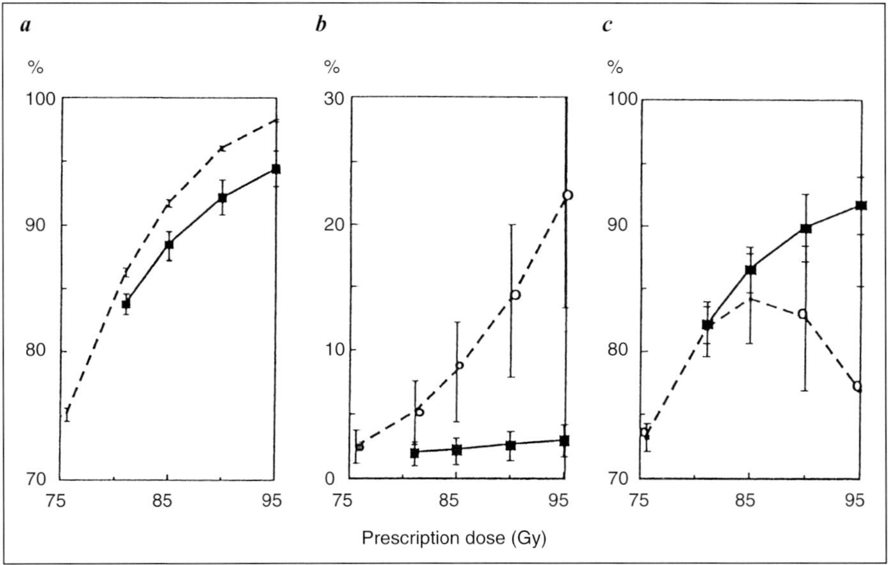


*Fig. 5.* TCP, NTCP and uncomplicated control, TCP $\times$ (1–NTCP), as a function of prescription dose for two types of plans. For the type II plan the rectal wall was shielded after 72 Gy. - - - = Type I; —— = type II.

there is very little increase in NTCP with dose. The probability of uncomplicated control continues to increase with prescription dose for type II plans.

*Figure of Merit for Treatment Decision*

Amols et al. [1] have introduced the concept of figure of merit (FM) represented by the equation:

$$FM = [1 - (1 - TCP)^a]^b \cdot (1 - NTCP^c)^d \tag{6}$$

where a, b, c and d are 4 positive adjustable variables. The FM has the following properties:

$$\begin{aligned} FM &= 1 \quad \text{when } TCP = 1 \text{ and } NTCP = 0 \\ FM &= 0 \quad \text{when } TCP = 0 \text{ and } NTCP = 1 \end{aligned} \tag{7}$$

FM increases monotonically with increasing TCP and FM decreases monotonically with increasing NTCP and in the special case:

$$a = b = c = d = 1 \tag{8}$$

the FM reduces to FM = TCP · (1–NTCP), the equation for uncomplicated
control. The parameters are unique for each physician, each disease site, and
perhaps for each individual patient. Amols et al. [1] refer to these parameters as
the physician parameters and they argue that the FM can be tailored to the pref-
erence of each physician, and/or to specific disease site or patient types. The
willingness to accept risk in exchange for possible cure is embedded in the four
parameters. It is quite possible too that well-informed patients may prefer to
have their own FM equation used (rather than the physician's) when selecting a
treatment course.

*Intensity-Modulated Plan Optimization*

Recently, computer-optimized intensity modulation treatment planning [2]
has provided a method of shaping the dose distribution to the shape of the PTV
while keeping the dose to critical structures within specified limits. Many
researchers have investigated the optimization of intensity distributions based
on biological objective functions. For example Wang et al. [25] have found that
the dose-based optimization produced satisfactory approximations of the
desired dose distributions for the treatment of the prostate, but not for the lung.
The application of the biology-based optimization produced significant
improvement for the lung plan, in terms of dose distributions, DVH and the values of biological indices. They argued that the differences in behavior of the
inverse technique for prostate and lung are most likely attributable to the differences in the tolerance doses of the neighboring normal tissues, the magnitudes
of the volume effect and tissue architecture. For prostate, the critical normal
structure tolerances relative to the prescription doses are high, and they exhibit
a small volume effect. For rectum, the parameters are $TD_{50}(1) = 80\,Gy$ and
$n = 0.12$. Hence, the objectives stated in terms of dose are achievable to a
greater degree and specifying the dose limits to the critical organs is adequate.
In contrast, for the lung the dose-limiting tissue surrounding the PTV is normal
lung tissue with much lower tolerance $[TD_{50}(1) = 24.5\,Gy]$ and a large volume
effect ($n = 0.87$). It is permissible to treat small lung volumes with high doses
as long as the volume irradiated with such doses is small. In these types of cases
TCP and NTCP based optimization can be helpful.

## Summary and Conclusion

Response of human tissues to ionizing radiation is a complex process. It is
influenced by many factors, such as use of chemotherapy drugs and underlying
diseases such as diabetes and/or lung emphysema. A phenomenological model
such as Lyman's is an attempt to predict the complication, for a variety of

tissues, in the absence of these factors. The use of the model requires the knowledge of the parameters to predict the response for a specific endpoint. Clinical response data are needed to determine these parameters. Emami et al. [6] have provided some data, based on pre-CT and pre-3-D information, for some of the most serious complications. Based on this information the parameters were determined [4]. However, to validate and further improve the predictive power of the model, improved clinical response data are needed.

With CT-based 3-D treatment planning systems the dose-volume information is routinely produced. Efforts by the radiation oncology community are needed to collect this information and correlate it with the clinical outcomes in a uniform and systematic way, not only for the most serious complications but also for less severe radiation-induced complications that are routinely considered in radiation therapy. Also, the information about the tissue response with underlying disease and drugs will be useful.

The use of NTCP for plan comparison is useful. However, the incorporation of TCP and NTCP for designing the plan is remarkable. A plan can be optimized for the best outcome for the patient. It is hoped that as the models and parameters are refined and predictive power of the model increases, better plans will be produced, significantly improving the therapeutic ratio.

## References


1   Amols HI, Zaider M, Hayes MK, Schieff PB: Physician/patient-driven risk assignment in radiation oncology: Reality or fancy? Int J Radiat Oncol Biol Phys 1997;38:455–461.
2   Bortfeld T, Kahler DL, Waldron TJ, Boyer AL: X-ray field compensation with multileaf collimators. Int J Radiat Oncol Biol Phys 1994;28:723–730.
3   Burman C, Happersett L, Kutcher G, et al: Use of radiobiological indices to guide dose escalation of the prostate cancer patients. Int J Radiat Oncol Biol Phys 1997;39(suppl):194.
4   Burman C, Kutcher GJ, Emami B, Goitein M: Fitting of normal tissue tolerance data to an analytic function. Int J Radiat Oncol Biol Phys 1991;21:123–135.
5   Drzymala RE, Mohan R, Brewster L, et al: Dose-volume histograms. Int J Radiat Oncol Biol Phys 1991;21:71–78.
6   Emami B, Lyman J, Brown A, et al: Tolerance of normal tissue to therapeutic irradiation. Int J Radiat Oncol Biol Phys 1991;21:109–122.
7   Goitein M: The probability of controlling an inhomogeneously irradiated tumour; in Report of the Working Group on the Evaluation of Treatment Planning for Particle Beam Radiotherapy. Bethesda, National Cancer Institute, 1987, pp 5.8.1–5.8.17.
8   Jackson A, Ten Haken RK, Robertson JM, Kessler ML, Kutcher GJ, Lawrence TS: Analysis of clinical complication data for radiation hepatitis using a parallel architecture model. Int J Radiat Oncol Biol Phys 1995;31:883–891.
9   Kutcher GJ: Quantitative plan evaluation; in Purdy J (ed): Advances in Radiation Oncology Physics. AAPM Monograph. New York, American Institute of Physics, 1992, No 19, pp 998–1021.
10  Kutcher GJ, Burman C: Calculation of complication probability factors for non-uniform normal tissue irradiation: The effective volume method. Int J Radiat Oncol Biol Phys 1989;16:1632–1630.
11  Lawrence TS, Ten Haken RK, Kessler ML, et al: The use of 3-D dose volume analysis to predict radiation hepatitis. Int J Radiat Oncol Biol Phys 1992;23:781–788.

12    Lyman JT: Complication probability – As assessed from dose-volume histograms. Radiat Res 1985;104:513–519.

13    Lyman JT, Wolbarst AB: Optimization of radiation therapy. III. A method of assessing complication probabilities from dose-volume histograms. Int J Radiat Oncol Biol Phys 1987;13: 103–109.

14    Lyman JT, Wolbarst AB: Optimization of radiation therapy. IV. A dose-volume histogram reduction algorithm. Int J Radiat Oncol Biol Phys 1989;17:433–436.

15    Martel MK, Sandler HM, Cornblath WT, et al: Dose-volume complication analysis for visual pathway structures of patients with advanced paranasal sinus tumours. Int J Radiat Oncol Biol Phys 1997;38:273–284.

16    Martel MK, Ten Haken RK, Hazuka MB, et al: Dose-volume histogram and 3-D treatment planning evaluation of patients with pneumonitis. Int J Radiat Oncol Biol Phys 1994;28:575–581.

17    Moiseenko V, Battista J, Van Dyk J: Normal tissue complication probabilities: Dependence on choice of biological model and dose-volume histogram reduction scheme. Int J Radiat Oncol Biol Phys 2000;46:983–993.

18    Munro P: Portal imaging technology: Past present and future. Semin Radiat Oncol 1995; 5:115–133.

19    Niemierko A: Reporting and analyzing dose distributions: A concept of equivalent uniform dose. Med Phys 1997;24:103–110.

20    Niemierko A, Goitein M: Implementation of a model for estimating tumor control probability for inhomogeneously irradiated tumor. Radiother Oncol 1993;29:140–147.

21    Photon Treatment Planning Collaborative Working Group: State-of-the-art of external photon beam radiation treatment planning. Int J Radiat Oncol Biol Phys 1991;21:9–23.

22    Smith AR, Purdy JA: Report of the Collaborative Working Group on the Evaluation of Treatment Planning for the External Photon Beam Radiotherapy. Int J Radiat Oncol Biol Phys 1991;21:1–265.

23    Schultheiss TE, Orton CG: Models in radiotherapy: Definition of decision criteria. Med Phys 1985;13:183–187.

24    Schultheiss TE, Orton CG, Peck RA: Models in radiotherapy: Volume effects. Med Phys 1983; 10:410–415.

25    Wang X-H, Mohan R, Jackson A, Leibel SA, Fuks Z, Ling CC: Optimization of intensity-modulated 3D conformal treatment plans based on biological indices. Radiother Oncol 1995;37: 140–152.

26    Zelefsky MZ, Leibel SA, Gaudin PB, et al: Dose escalation with three-dimensional conformal radiation therapy affects the outcome in prostate cancer. Int J Radiat Oncol Biol Phys 1998;41: 491–500.

Chandra M. Burman, Department of Medical Physics,
Memorial Sloan-Kettering Cancer Center, 1275 York Avenue, New York, NY 10021 (USA)
Tel. +1 212 639 8787, Fax +1 212 717 3258, E-Mail burmanc@mskcc.org

Dörr W, Engenhart-Cabillic R, Zimmermann JS (eds): Normal Tissue Reactions in Radiotherapy and Oncology. Front Radiat Ther Oncol. Basel, Karger, 2002, vol 37, pp 163–173

# Intensity Modulation Techniques for Improvement of Normal Tissue Tolerance

*W. De Neve, F. Claus, W. Duthoy, G. De Meerleer, C. De Wagter*

Division of Radiotherapy, Ghent University Hospital, Ghent, Belgium

In the 1980s, Brahme and colleagues [1, 16] computed fluence patterns that were useful to form homogeneous concave dose distributions by arc therapy. The interest in using these fluence patterns – in what was later called intensity modulated beams – was also shown for multiple static beams. Organs at risk (OARs) surrounded by tumor could now be spared. For a given tumor prescription dose, intensity-modulated radiotherapy (IMRT) can be exploited to improve normal tissue tolerance by physical selectivity, i.e. the physical dose is larger in the tumor than in invaginating normal tissues. With fractionation, the therapeutic gain is greater than expected from the degree of physical selectivity. Indeed, by lowering the dose to OARs for each fraction, physical selectivity is complemented by biological (fractionation) selectivity when the ratios of $\alpha/\beta$-values between tumor and OARs have a value larger than 1. Some IMRT applications using multiple convergent beams or arc therapy are characterized by spreading out the dose over large volumes. Regions receiving a dose per fraction below 0.5 Gy occur and the issue of low-dose hypersensitivity may have to be considered. The aim of this paper is to illustrate the use of IMRT for the improvement of normal tissue tolerance by exploiting physical and biological selectivity. The potential drawback of low-dose hypersensitivity and the possible precautions to minimize the risk of its toxicity are also discussed. IMRT treatments selected from a database of patients treated for head and neck cancer are taken to illustrate the arguments.

| Site | Patient number | Reirradiation |
|---|---|---|
| Nasopharynx | 8 | 3/8 |
| Oropharynx | 3 | 2/3 |
| Hypopharynx | 2 | 1/2 |
| Larynx | 4 | 3/4 |
| Oral cavity | 2 | 1/2 |
| Parotid | 1 | 1/1 |
| Thyroid gland | 4 | 0/4 |
| Paranasal sinus | 8 | 0/8 |

## Materials and Methods

### Patient Treatment

Between September 1, 1996 and October 31, 1999, 32 patients with cancer in the head and neck region were treated using segmental IMRT. Table 1 shows the distribution of patients according to the site of the primary. Eleven of these patients were reirradiated for relapse or metachronous tumor in a previously irradiated region (table 1). Permission from the hospital's Ethical Committee was obtained to use IMRT for reirradiation in symptomatic patients with inoperable disease. IMRT dose distributions were required that could achieve a large difference between the dose delivered to the malignancy and the dose delivered to anatomical structures that had previously been irradiated at doses close to tolerance. Thus the key mechanism that we aimed for was physical selectivity. Reirradiation by IMRT was performed from September 1996.

From August 1998 till the end of October 1999, 8 patients received IMRT for paranasal sinus cancer. In all patients, regions of the clinical target volume were at close distance (within 0–3 mm) from optical structures and pathways. As a result, the planning target volumes (PTVs) intersected optical structures (retina) and/or pathways (optic nerves, optic chiasm). In all patients, conventional radiotherapy techniques would have involved the delivery of total doses exceeding 60 Gy to optical OARs which would be more than the tolerance levels of these structures. In standard radiotherapy practice unilateral blindness is a feared complication of radiotherapy in such cases. The treatment planning often involves the sinister choice of 'which eye to sacrifice'. The challenge for IMRT is to save binocular vision. The key mechanism is combined physical and biological selectivity. Follow-up being too short, we aim to show that IMRT allows us to achieve dose distributions that are suitable to pursue this mechanism.

In selected patients with pharyngeal cancer, parotid sparing was performed. The treatment of one of these patients is taken as an example to discuss the potential adverse effects of low-dose hypersensitivity when using forms of IMRT that lead to large areas that are irradiated at low doses per fraction. The patient was an 8-year-old boy diagnosed with a nasopharyngeal embryonal rhabdomyosarcoma. The patient was treated by induction chemotherapy (ifosfamide, vincristine, Adriamycin: 3 cycles) followed by radiotherapy (56 Gy in 2-Gy fractions prescribed to PTV based on the pretherapeutic extension of the gross tumor volume). The first 5 fractions were delivered by a 3D conformation technique while the remaining 23 fractions were delivered by IMRT. The 3D conformation technique did not allow us to spare a parotid gland but was applied to enable us to start radiotherapy in due time, i.e. according to the protocol guidelines. After 5 fractions, the IMRT plan and its dosimetric verification were finished. This practice of starting with conventional radiotherapy until a sophisticated IMRT plan is finished and preclinically verified is not uncommon. To discuss the issue of low-dose hypersensitivity, a number of assumptions are made. First, it is assumed that low-dose hypersensitivity occurs in the parotid gland, that it occurs at fraction doses of 50 cGy/fraction or lower and that for 45 cGy/fraction, the biological equivalent dose (BED) is 2.0 times the physical dose. We also assume that a rival IMRT plan is available in which the right parotid would receive a mean and homogeneous dose of 45 cGy/fraction. We want to stress the fact that low-dose hypersensitivity has recently been reported for parotid glands [15]. However, the exact quantitative hypersensitivity data are not known to us. Therefore, the proposed dose-modifying factor of 2.0 for a fraction dose of 0.45 Gy and the critical dose of 0.5 Gy below which hypersensitivity would occur have to be considered in the context of a further theoretical discussion only.

*Equipment*

Software distributed by Sherouse (GRATIS® [19]) was used as a platform for IMRT planning. Inhouse developed tools were added to translate field edges to multileaf collimator settings [24], to perform beam segmentation for IMRT [7], to optimize beam weights [3, 8], and to optimize collimator angles and individual multileaf collimator (MLC) leaf positions [De Gersem, unpubl. data]. Treatments were delivered by means of 6-MV photons of an SL20-MLCi or an SL25-MLCi (Elekta, Crawley, UK) linear accelerator equipped with a dynamic multileaf collimator (DMLC). The DMLC control software was described previously [6]. An extensive system for preclinical dosimetric verification and clinical quality control has been developed [2, 9, 23].

## Results

*Reirradiation*

Plans were constructed with the aim to achieve the maximum possible sparing of previously irradiated OARs such as spinal cord, brainstem and mandibular bone. Figure 1 shows a typical plan. Maximal sparing of the brainstem and the spinal cord results in a dose distribution showing a low-dose tube matching brainstem and spinal cord. In any transverse direction away from brainstem and spinal cord, a sharp dose gradient exists. Patient setup and immobilization are critical. There are two reasons for dose inhomogeneity in the PTV. First, underdosage is observed where the minimal distance between PTV and spinal cord or brainstem is of the order of 1 cm or less, i.e. the same order of magnitude as the beam penumbra. The beam penumbra sharpness determines the maximum dose gradient that can be achieved around an invaginating structure. Second, when a concave dose distribution must be created, with the aim to deliver a zero dose to the invaginating structure, the maximum dose

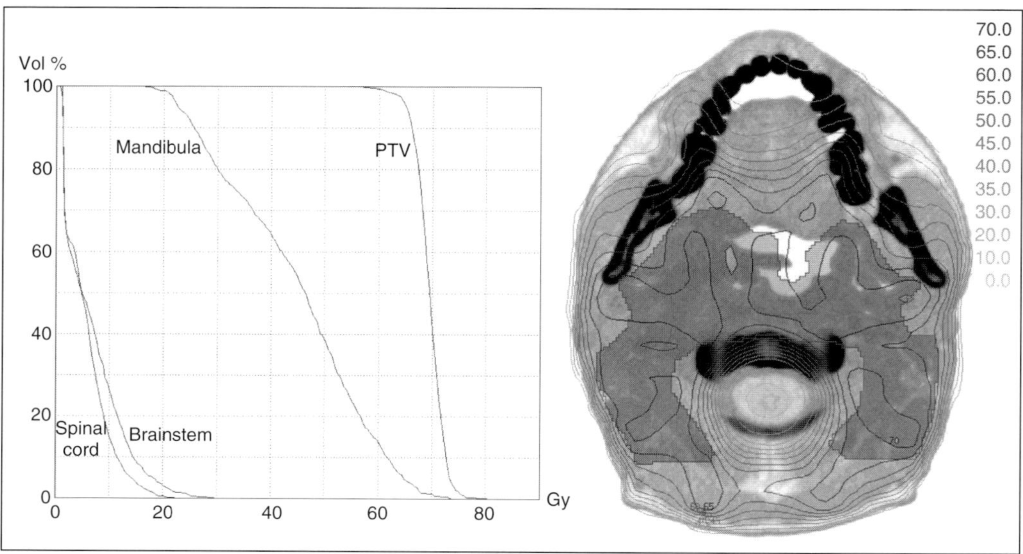


*Fig. 1.* Dose-volume histogram and transverse isodose line plot of IMRT for a large horseshoe-shaped relapse from a nasopharyngeal cancer. The PTV is shaded on the transverse plot. A dose gradient is shown, running from 20 to 60 Gy over a distance of about 1 cm in the centrifugal direction from the spinal cord. The dose to two thirds of the mandible is lower than 55 Gy.

homogeneity that can be achieved depends on the number of incident static beams. If the number of beams is n, the dose homogeneity greater than $(n - 1)/n$ is difficult to obtain. Table 2 shows the location of the primary tumor (column 1), the dose delivered initially (column 2) and for reirradiation (column 3), the response to treatment (column 4) and the events observed during follow-up (column 5). For 8 of 10 evaluable patients palliation was achieved. Median duration of response was 9 months (fig. 2). Median survival calculated from the onset of reirradiation was 15 months (fig. 2). The cause of death was intercurrent (intestinal hemorrhage) in 1 patient and related to disease progression in 5 patients. Subcutaneous fibrosis was reported in 7 patients, temporomandibular joint impairment and laryngeal edema each in 1 patient. In spite of cumulative doses as high as 136 Gy, no cases of cranial nerve palsy, arterial rupture or bone necrosis were observed.

### Intentionally Inhomogeneous Dose Distributions

In paranasal sinus cancer (especially ethmoid cancer), tumor edges adjacent to one or more of the numerous OARs (eyes, optic nerves and chiasm, brainstem, frontal lobes) are common. Conformal avoidance techniques of

| Site | Initial dose Gy | IMRT Gy | Response | Follow-up/ outcome |
|---|---|---|---|---|
| Nasopharynx | 66 | 60 | CR | 7/no evidence of disease |
| Nasopharynx | 66 | 70 | CR | 8/relapse |
| Nasopharynx | 64.8 | 70 | CR | 7/no evidence of disease |
| Parotid gland | 64 | 60 | PR | 14/no evidence of disease |
| Oropharynx | 70 | 60 | PR | 23/relapse |
| Hypopharynx | 64.8 | 60 | PR | 3/relapse |
| Oropharynx | 66 | 70 | CR | 10/relapse |
| Oral cavity | 50 | 66 | PD | 0/relapse |
| Larynx | 66 | 60 | PR | 2/relapse |
| Larynx | 70 | 44* | ? | 4/*stopped RT |
| Larynx | 70 | 40* | ? | 0/*died during RT |

Follow-up is given in months. CR = Complete response; PR = partial response; PD = persistent disease; RT = radiotherapy.

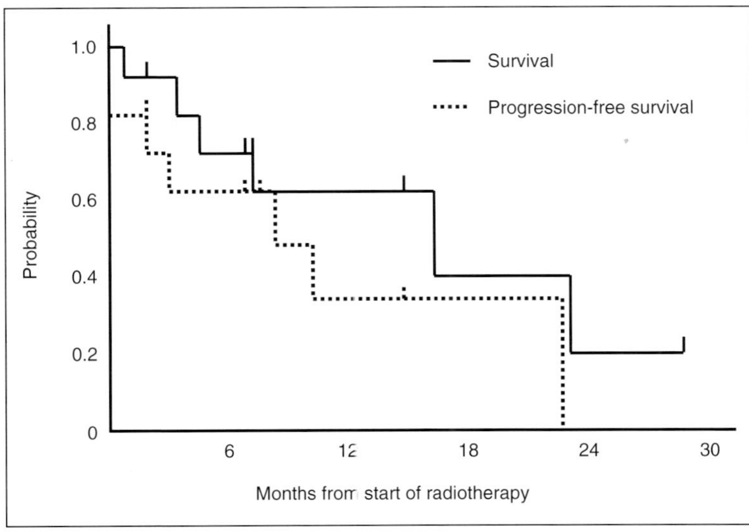


*Fig. 2.* Probability of survival and progression-free survival for patients (n = 11) reirradiated for head and neck cancer. Date of analysis: January 15, 2000.

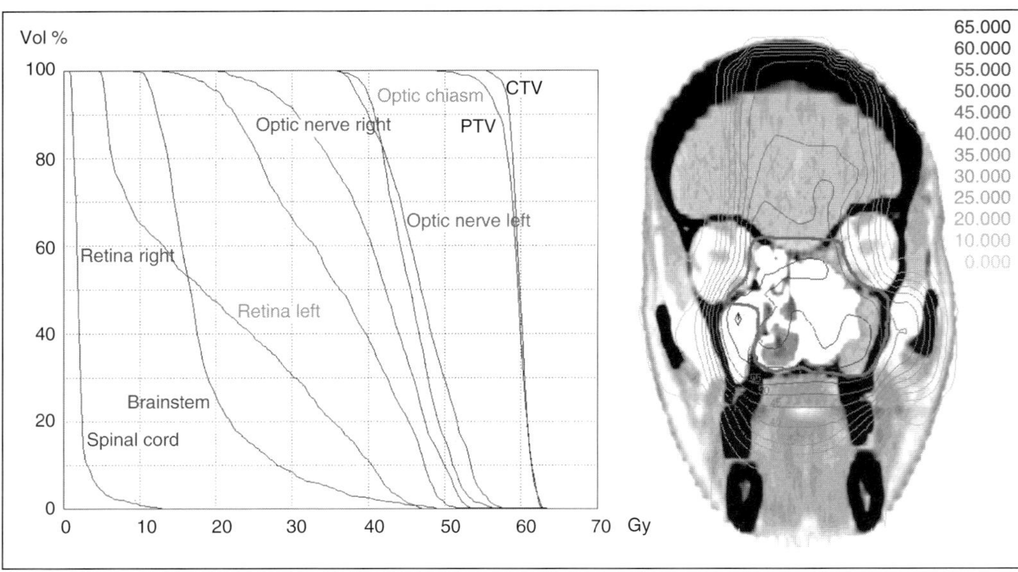


***Fig. 3.*** Dose-volume histogram and coronal isodose line plot of postoperative IMRT for ethmoid sinus cancer. PTV is shown as solid gray line on the coronal plot. PTV prescription dose was 30 × 2.0 Gy. Left and right optic nerves pass through dose gradient. Doses above 50 Gy in optic nerves and chiasm are in regions that are intersected by the PTV [not shown on the coronal plot but consistent with the difference in dose minimum between PTV and clinical target volume (CTV)].

small-size OARs must take into account motion and setup uncertainty, not only of the tumor but also of the OARs. Indeed, when creating sharply increasing dose gradients in the vicinity of structures like the optic nerves (4 mm diameter) or optic chiasm (7 mm diameter), setup errors may expose the full thickness of the structure above tolerance. Applying margins for motion and setup uncertainty to clinical target volume as well as to OARs adds safety. The resulting PTV often intersects the expanded OARs. The maximum dose to the region of intersection is determined by the tolerance of the respective OARs, while the dose to the nonintersecting PTV part can be prescribed to a higher value. IMRT allows to irradiate with an inhomogeneous dose distribution so that at each fraction the PTV portion that intersects with or is close to the OAR(s) is slightly underdosed. If the tumor has a larger $\alpha/\beta$ than the OARs (often the case if the OAR is nerve tissue), a radiobiological advantage can be exploited by the smaller fraction size at the OAR. As a result of the smaller fraction size to OARs, the tolerance dose is increased. In figure 3, the dose gradient between optic nerves, chiasm and tumor runs through a small portion of the tumor. In

none of the 8 patients treated between February 1999 and October 31, 1999, did
blindness develop. Duration of follow-up, however, is insufficient.

*Low-Dose Hypersensitivity: An Unforeseen Obstacle for IMRT?*

Avoiding xerostomia became another issue in IMRT of head and neck cancer, typically for the pharyngeal site. Dose/volume/function relationships in the parotid glands were found to be characterized by dose and volume thresholds, steep dose/response relationships when the thresholds were reached, and a high volume dependence in normal tissue complication probability models. Eisbruch et al. [10] concluded that a mean dose of $\leq 26$ Gy to the parotid gland should be a planning goal if substantial sparing of the gland function is desired. In our experience, a free space of 1–1.5 cm between the PTV and the parotid gland was sufficient to achieve this planning goal for a PTV prescription dose up to 70 Gy.

For the irradiation of the nasopharyngeal rhabdomyosarcoma, we assumed that a theoretical plan was developed that spared the right parotid gland. At a PTV dose prescription of 2 Gy/fraction, the right parotid mean physical dose would be 0.45 Gy/fraction. Following the assumption of low-dose hypersensitivity as described in the Materials and Methods section, the corresponding BED would be 0.9 Gy/fraction. For a PTV dose prescription of $28 \times 2.0$ Gy, the BED to the right parotid would be $28 \times 0.9 = 25.2$ Gy and thus below the planning goal (i.e. a total mean dose below 26 Gy, BED $\simeq 34$ Gy) to spare the parotid gland. Since there was no parotid sparing during the initial 5 fractions, a dose of 10 Gy (physical) or 16.7 Gy (BED, $\alpha/\beta = 3$ Gy) was accumulated. Applying the theoretical plan after 5 fractions would add a BED of $23 \times 0.9 = 20.7$ Gy. The BED for the total treatment would be 37.4 Gy, thus violating the planning goal, while the physical total dose would be $10 + 23 \times 0.45 = 20.35$ Gy mean dose to the right parotid, i.e. in compliance with Eisbruch's planning goal but misleading.

## Discussion

For patients with inoperable locoregional relapse or new primary after high-dose radiotherapy for head and neck cancer, the prognosis is poor. In the absence of distant metastases, cure can be attempted by reirradiation. Most of the patients referred to our center suffer from advanced relapses. IMRT offers a sufficiently large window of possibilities to perform reirradiation for large tumors that encompass OARs with limited remaining tolerance to radiation. The dose to such OARs can be limited to values that are close to the lowest values physically achievable with current photon technology. When different OARs are inside the volume of reirradiation, priorities must be assigned as a function of the severity and likelihood of toxicity. Data about normal tissue tolerance for reirradiation are scarce. In practice, plan optimization aims at the delivery of the lowest possible dose to OARs such as brainstem and spinal cord, since complications would be invalidating or lethal. Our preliminary data suggest that IMRT is a safe alternative to palliative care in advanced inoperable patients. Palliation of symptoms by induction of tumor response is the preferred and logical approach and was achieved in 8/9 evaluable patients. The median duration of response was 8 months. We plan to apply reirradiation for less advanced cases.

The prime example of successful application of intentionally inhomogeneous dose distributions is the prostate [17]. The rectal wall, in close vicinity of the posterior portion of the prostate and seminal vesicles, is dose-limiting. When accounting for motion, the resulting PTV intersects the anterior rectal wall. The maximal homogeneous dose that can be given to the PTV is then directly determined by the intersected portion of the rectal wall. Escalating the dose above the rectal tolerance level can only be attempted safely to the non-intersecting PTV part. Although this strategy has not yet been shown conclusively to produce higher cure rates, it looks promising and does seem to limit side effects [11]. In cancer of the paranasal sinuses, a similar problem occurs. Cancers at this site(s) are often diagnosed in an advanced stage with invasion of one or both orbits, the anterior and middle fossae of the skull. Tumors in close vicinity or even displacing the optical pathways are often seen. Data regarding tolerance doses of optic nerves, chiasm and retinae with the use of segmental IMRT for advanced paranasal sinus tumors were published by Martel et al. [18]. She concluded that a maximal dose of 60 Gy in 1.8–2.0 Gy/fraction to the optic nerves is a suitable planning goal if visual function should be preserved. Of the optic pathway structures, the optic nerves seemed to have the lowest tolerance doses. For a PTV prescription dose of 70 Gy in 2.0 Gy/fraction, an underdosage of 20% to the optic tract (e.g. 1.6 Gy/fraction) would result in a physical dose of 56 Gy and a BED of 51.5 Gy at 2 Gy/fraction ($\alpha/\beta = 3$ for optic pathway structures). Accurate control of the maximal dose to optical pathways in the vicinity of sharp dose gradients (running from low in the optic pathways to high in the PTV) is a challenge even for modern high-precision techniques. The penumbras of the best collimation techniques (3–4 mm for the distance between the 20 and the 80% isodose) as well as the leaf width of minimultileaf collimators are about as large as the diameter of the optic nerves. The sharpest gradients inside an IMRT dose distribution are created by segment edges and the penumbra limits the maximal gradient steepness. A gradient running from 1.6 Gy to 2.0 Gy can be created over a distance of 4 mm [De Wagter, unpubl. data]. It is assumed that the 1.6-Gy isodose surface is positioned at the edge of the optic nerve. Noninvasive patient setup and immobilization in the head and neck region are characterized by random setup errors around 2 mm [5]. With a systematic error of zero, the maximal dose to the optic nerve would be 2.0 Gy for 2.5% of the setups. In clinical practice, IMRT may create an underdosed tube at the optic nerves. Setup errors in any direction would overdose the optic nerves. Although random errors tend to overdose different regions for different fractions, systematic errors would overdose the same region repeatedly. For radiotherapy of paranasal sinus tumors, we reach the limits that are achievable with the present photon technology and patient setup methods.

During the last decade, hypersensitivity of exponentially growing cells exposed to radiation doses below 0.5 Gy has been observed [21]. As underlying hypothesis, it was suggested that a dose of 0.5 Gy or less is not sufficient to trigger an inducible repair mechanism. If this effect were present in irradiated cells it would be of considerable clinical significance. In pulse dose brachytherapy, hyperfractionated radiotherapy, multibeam conformal therapy or tomotherapy, critical normal tissues often receive doses in the range of 0.5 Gy/fraction or lower. The evaluation of rival plans based on physical doses and comparisons using the LQ model might lead to erroneous selection. Optimization algorithms involving normal tissue complication probability computations based on the LQ model [3, 4] might lead to misleading predictions. Normal tissues showing low-dose hypersensitivity would impact on the choice of beam directions, number of beams and intensity profiles. The much debated issue of the best beam assembly in IMRT would become even more complicated and the arguments in favor of tumor-site-dependent class solutions would be strengthened. Clinical IMRT applications using hyperfractionation and two-phase plans (non-IMRT followed by IMRT) would not be favored. Hypofractionation could be advantageous.

Most publications involve observations in the laboratory. Not all observations using cell lines support that, in fractionated schedules, low-dose hypersensitivity would lead to surviving fractions that would be ill predicted by the well-documented utility of the LQ approach for estimating isoeffect doses for alternative fractionation schemes [20]. In mice, low-dose hypersensitivity was shown for acute skin damage and late kidney effects [13, 14]. Clinical data are scarce. Hamilton et al. [12] found that the LQ model significantly underpredicted peak skin erythema values at doses of less than 1.5 Gy/fraction. Turesson et al. [22] also found evidence of low-dose hypersensitivity in human skin exposed to fractionated radiotherapy. Hypersensitivity in cell kill for doses up to 0.2–0.4 Gy/fraction was followed by less cell kill between 0.45 and 1.1 Gy/fraction. Hypersensitivity did not seem to disappear as a function of cumulative doses and could be observed after 15 and 20 fractions at 3 and 4 weeks of radiotherapy, respectively. Lambin reported low-dose hypersensitivity in parotid glands [15]. Turesson and Joiner [21] pointed out that low-dose hypersensitivity is not encountered in all cell types or organs. Neither the list of organs nor the quantitative degree of hypersensitivity is known. Nevertheless, low-dose hypersensitivity has to be taken into account when planning IMRT.

## Acknowledgments

The project Conformal Radiotherapy Ghent University Hospital is funded by the Belgische Federatie tegen Kanker, by a gift from Omer Roelandt-Deleenheer and by grants of the Fonds voor Wetenschappelijk Onderzoek Vlaanderen, projects 98/121, G.0049.98 and G.0039.97, the Centrum voor Studie en Behandeling van Gezwelziekten and the Sportvereniging tegen Kanker. F. Claus and W. Duthoy are research assistants of the FWO. G. De Meerleer is postdoctoral fellow of the FWO. R-UZG is a member of the Elekta IMRT Consortium. G. De Smet is acknowledged for logistic support.

## References

1   Brahme A, Roos JE, Lax I: Solution of an integral equation encountered in rotation therapy. Phys Med Biol 1982;27:1221–1229.
2   De Deene Y, De Wagter C, Van Duyse B, et al: Validation of MR-based polymer gel dosimetry as a preclinical three-dimensional verification tool in conformal radiotherapy. Magn Reson Med 2000;43:116–125.
3   De Gersem W, Derycke S, Colle C, De Wagter C, De Neve W: Inhomogeneous target-dose distributions: A dimension more for optimization? Int J Radiat Oncol Biol Phys 1999;44:461–468.
4   De Gersem W, Derycke S, Colle C, De Wagter C, De Neve W: Optimization of beam weights in conformal radiotherapy planning of stage III NSCLC: Effect on therapeutic ratio. Int J Radiat Oncol Biol Phys 2000;47:255–260.
5   De Neve W, De Gersem W, Fortan L, Van den Heuvel F: Clinical implementation of electronic portal imaging: Correction strategies and set-up errors. Bull Cancer Radiother 1996;83:401–405.
6   De Neve W, De Gersem W, Derycke S, et al: Clinical delivery of intensity modulated conformal radiotherapy for relapsed or second-primary head and neck cancer using a multileaf collimator with dynamic control. Radiother Oncol 1999;50:301–314.
7   Derycke S, Van Duyse B, De Wagter C, De Neve W: Non-coplanar beam intensity modulation allows large dose-escalation in stage III lung cancer. Radiother Oncol 1997;45:253–261.
8   De Wagter C, Colle CO, Fortan LG, et al: 3D-conformal intensity modulated radiotherapy planning: Interactive optimization by constrained matrix inversion. Radiother Oncol 1998;47:69–76.
9   De Wagter C, Martens C, De Deene Y, De Gersem W, Van Duyse B, De Neve W: Dosimetric verification of intensity modulation executed by a conventional accelerator equipped with a multileaf collimator. Cancer Radiother 1999;3(suppl 1):171–182.
10  Eisbruch A, Ten Haken RK, Kim HM, Marsch LH, Ship JA: Dose, volume and function relationships in parotid salivary glands following conformal and intensity-modulated irradiation of head and neck cancer. Int J Radiat Oncol Biol Phys 1999;45:577–588.
11  Garnick MB, Fair WR: Combating prostate cancer. Sci Am, December 1998, pp 45–53.
12  Hamilton CS, Denham JW, O'Brien M, et al: Underprediction of human skin erythema at low doses per fraction by the linear quadratic model. Radiother Oncol 1996;40:23–30.
13  Joiner MC, Denekamp J, Maughan RL: The use of 'top-up' experiments to investigate the effect of very small doses per fraction in mouse skin. Int J Radiat Biol 1986;49:565–580.
14  Joiner MC, Johns H: Renal damage in the mouse: The response to very small doses per fraction. Radiat Res 1988;114:385–398.
15  Lambin P, Van Mellaert L, Van Acker F, et al: Direct evidence of hypersensitivity and induced radioresistance in human systems. Int J Radiat Oncol Biol Phys 2000;46:692.

16   Lax I, Brahme A: Rotation therapy using a novel high-gradient filter. Radiology 1982;145:473–478.

17   Ling CC, Burman C, Chui C, et al: Conformal radiation treatment of prostate cancer using inversely-planned intensity-modulated photon beams produced with dynamic multileaf collimation. Int J Radiat Oncol Biol Phys 1996;35:721–730.

18   Martel MK, Sandler HM, Cornblath WT, et al: Dose-volume complication analysis for visual pathway structures of patients with advanced paranasal sinus tumors. Int J Radiat Oncol Biol Phys 1997;38:273–284.

19   Sherouse GW, Thorn J, Novins K, Margolese-Malin L, Mosher C: A portable 3D radiotherapy treatment design system. Med Phys 1989;16:466.

20   Smith LG, Miller RC, Richards M, Brenner DJ, Hall EJ: Investigation of hypersensitivity to fractionated low-dose radiation exposure. Int J Radiat Oncol Biol Phys 1999;45:187–191.

21   Turesson I, Joiner MC: Clinical evidence of hypersensitivity to low doses in radiotherapy. Radiother Oncol: 1996;40:1–3.

22   Turesson I, Flogegard M, Johansson KA, Svensson P, Wahlgren T, Nyman J: Use of in-situ assays and molecular markers to measure individual human tissue response to radiotherapy: Evidence of low dose hypersensitivity and proliferation dependent radiosensitivity. Int J Radiat Oncol Biol Phys 2000;46:737.

23   Vakaet LAML, Bate M-T, Fortan LG, De Wagter C, De Neve WJ: Off-line verification of the day-to-day three-dimensional table position variation for radiation treatments of the head and neck region using an immobilization mask. Radiother Oncol 1998;47:49–52.

24   Van Duyse B, Colle C, De Wagter C, De Neve W: Modification of a 3D-planning system for use with a multileaf collimator; in Conformal Radiotherapy: Physics, Treatment Planning and Verification. Ghent, Belgian Hospital Physicist Association, 1995, pp 39–48.

Wilfried De Neve, Division of Radiotherapy, Ghent University Hospital,
De Pintelaan 185, B–9000 Ghent (Belgium)
Tel. +32 9 24030 15, Fax +32 9 24030 40, E-Mail wilfried@krtkgl.rug.ac.be

Dörr W, Engenhart-Cabillic R, Zimmermann JS (eds): Normal Tissue Reactions in Radiotherapy and Oncology. Front Radiat Ther Oncol. Basel, Karger, 2002, vol 37, pp 174–184

..........................

# How the Game Is Played – Challenge between Therapeutic Benefit and Acute Toxicity in Fractionated Radiotherapy

*Bogusław Maciejewski, Leszek Miszczyk, Rafał Tarnawski, Krzysztof Składowski*

Department of Radiotherapy, Cancer Center MSC Institute, Gliwice, Poland

Experimental and many clinical data show convincing evidence that extension of overall treatment time (OTT) reduces the efficacy of radiotherapy, mainly due to accelerated tumor clonogen repopulation. Therefore, reduction in OTT may likely increase the probability of locoregional control (LRC) for a given total dose. On the other hand, better understanding of differences in the sensitivity to changes in dose per fraction between acutely (and most epithelial cancers alike) and late-responding tissues have led to an increase in these differences by delivering fractions smaller than 2.0 Gy more often than once a day. These findings became a basic rationale to design various altered, i.e. accelerated (AF) or hyperfractionation (HF) schedules. Hybrids of AF and HF, as the result of complex alterations of fractionation parameters, have also been tested to enhance the therapeutic ratio by improving tumor LRC without increasing late toxicity.

### How to Achieve the Therapeutic Gain?

Altered fractionation schedules have been the subject of extensive clinical studies, especially during the last decade, and some of the trials show a benefit for some head and neck cancer sites and stages [1, 2, 4, 5, 9, 11]. Table 1 shows the results of the most often cited trials in recent years.

For the first time the EORTC 22791 trial provided evidence for the advantage of pure HF, given in two small (1.15 Gy) daily fractions, over once-a-day treatment, and produced 18% increase in the LRC [4]. This gain was mostly observed in patients with larger tumors ($T_3$) regardless of nodal status. Besides an extra 10.5 Gy delivered in the HF arm (in fact, the equivalent $NTD_2$[1] was much smaller, equal to 4.8 Gy), weekly acceleration of dose intensity was also higher by a factor of 1.15 Gy in the HF arm than in control. The observed advantage of pure HF over conventional fractionation may result from a favorable combination of several radiobiological reasons, as it was suggested by Horiot et al. [4].

The AF trials consistently report a better LRC in favor of accelerated treatments, although those with the lowest total doses (CHART, PMH) showed borderline advantages (5% increase in the LRC). The EORTC 22851 trial [5] with the OTT reduced from 7 to 5 weeks without changing the total dose yielded a significant 13% gain in LRC. In turn in the CHART [2], a large reduction of the total dose associated with drastic shortening of the OTT seems to neutralize its potential benefit. In both, the CHART and EORTC 22851 the LRC gain was more likely observed in larger tumors, whereas the reverse was seen in the PMH study [1], where for tumors smaller than 4 cm a 1% gain in LRC for each 1% increase in total dose was noted. The best improvement was observed for hypopharynx, whereas in CHART it was for larynx tumors.

It becomes obvious that these three AF trials investigated markedly different fractionation schedules used for a wide variety of tumor sites and stages. It seems almost impossible to make a useful comparison and to separate the effect of dose fractionation from that of treatment time. Furthermore, other differences between centers, i.e. technical and dosimetric factors, should not be ignored, as they may flatten the dose-response curves and reduce the real gain from HF/AF schedules. Although all these trials provide clear evidence (fig. 1a) that the treatment time is the major factor determining the rate of gain (or failure), it does not seem to be the only one.

In two other trials (DAHANCA, CAIR), including similar tumor sites and stages, the OTT was the only variable [9, 11]. Avoiding uncompensated 3-week splits, DAHANCA-5 gave a 20% gain in the LRC compared with DAHANCA-2. A further 1 week reduction of the OTT to 5.5 weeks (DAHANCA-7) by treating on 6 days a week produced an extra gain of 10% in LRC (even 18% of LTC). CAIR goes one step further by giving a 7-day treatment and reducing the OTT from 7 to 5 weeks. It brought a 45% gain in LRC, giving an average 3.2% improvement in LRC per 1-day shortening.

These two trials show that by extending the treatment to 1 or even 2 days of the weekend, dose intensity also increases, which gives a higher and faster accumulation of the 'effective' dose delivered to the tumor because less of the dose

---

[1] $NTD_2$ is the normalized total dose if given in 2.0-Gy fractions, calculated using the L-Q equation with an $\alpha/\beta$ value of 10 Gy. Any $\alpha/\beta$ value above 10 Gy (10–25 Gy) gives only 1–5% change in the calculated $NTD_2$.

***Table 1.*** LRC and severe acute mucosal reactions (CM-IV and CLE) for different altered fractionation schedules

| Type and scheme (TD/n/OTT) | Authors | Fractionation | | Tumor response | | | Acute effects | | | |
|---|---|---|---|---|---|---|---|---|---|---|
| | | Interval between fractions | Treatment days/week | $NTD_{2.0}$ ($\alpha/\beta = 10$) Gy | $\Delta NTD_{eff}$ between two arms Gy | LTC (3–5 years) % | $AD_1$ | $RAD_2$ | CM % | CLE % |
| (a) AF, t.i.d. 59.4 Gy/33/25 days | Lamb et al. [8] | 4 | 3 | 58.4 | | 47 (all) | 16.2 | 1.62 | 80 | – |
| (b) SAF, e.i.d. 72 Gy/40/26 days | Nguyen et al. [10] | 2 | 5 | 65.4 | | 40 (2 years) | 36.0 | 2.23 | 100 | 80 |
| (c) AF, t.i.d. 54 Gy/27/12 days | Peracchia and Salti [12] | 3–4 | 5 | 54.0 | | – | 30.0 | 2.04 | 100 | 55 |
| (d) HF, b.i.d. PMH, 58 Gy/40/28 days vs. 51 Gy/20/28 days | Cummings et al. [1] | 6 | 5 | 55.3 52.9 | 2.4 | 45 40 | 14.5 12.75 | 1.45 1.28 | 63 40 | – – |
| (e) SAF, t.i.d. – EORTC 22851 72 Gy/45/35 days vs. 70 Gy/35/49 days | Horiot et al. [5] | 4 | 5 | 69.6 (61.4)[a] | 8.2 | 59 46 | 24.0 10.0 | 1.44 1.0 | 70 34 | 2 (?) – |
| (f) HF, b.i.d. – EORTC 22791 80.5 Gy/70/49 days 70 Gy/35/49 days | Horiot et al. [4] | 4–6 | 5 | 74.8 70.0 | 4.8 | 56 38 | 11.5 10.0 | 1.15 1.0 | 67 49 | |

| | | | | TD | ΔNTD$_{eff}$ | | AD$_1$ | RAD$_2$ | LTC | |
|---|---|---|---|---|---|---|---|---|---|---|
| (g) BAF, q.d.→b.i.d.<br>70 Gy/35/39 days | Kaanders<br>et al. [6] | 6 | 5 | 70.0 | | – | 10.0 | 1.0 | 90 | |
| (h) DAHANCA-2, 5, 7<br>D-2, 66 Gy/33/66 days (SF) | Overgaard<br>et al. [11] | | 5 | 47.8[b] | –18.2 | 32 (33%) | 10.0 | 1.0 | | |
| D-5, 66 Gy/33/45 days (CF) | | | 5 | 66.0 | | 52 (61%) | 10.0 | 1.0 | | |
| D-7 66 Gy/33/38 days (AF) | | | 6 | 72.2[b] | +6.2 | 62 (77%) | 12.0 | 1.2 | 73 | |
| (i) EAF, b.i.d.<br>76 Gy/54/35 days | Harari [3] | 6 | 6 | 72.0 | | – | 14.0 | 1.4 | 100 | |
| (k) AHF, t.i.d. – CHART<br>54 Gy/36/12 days vs. | Dische<br>et al. [2] | 6 | 7 | 51.75 | 1.75 | 45 | 31.5 | 2.7 | 73 | |
| 66 Gy/33/42 days | | | 5 | 50.0[c] | | 40 | 10.0 | 1.0 | 43 | |
| (l) AF, q.d. – CAIR<br>70 Gy/35/35 days vs. | Maciejewski<br>et al. [9] | | 7 | 76.0[d] | 14.4 | 82 | 14.0(12.6) | 1.4 | 62 | 22 (0%) |
| 70 Gy/35/49 days | | | 5 | (61.4)[a] | | 37 | 10.0 | 1.0 | 26 | |

LTC = Local tumor control; TD = total dose; n = number of fractions; NTD$_{2.0}$ = normalized total dose if given in 2-Gy fractions; ΔNTD$_{eff}$ = increase in the NTD in study aim in relation to control arm; SAF = split AF; BAF = boost AF; EAF = escalated AF; AHF = accelerated HF; AD$_1$ = accumulated dose in the 1st week; RAD$_2$ = accumulated dose in week 2 relative to two conventional AD$_2$ = 20 Gy; CF = conventional fractionation.

[a] NTD normalized to OTT of 35 days (repopulation rate = 0.6 Gy/day).
[b] NTD normalized to OTT of 45 days.
[c] NTD normalized to OTT of 12 days.
[d] NTD normalized to 5-day treatment.

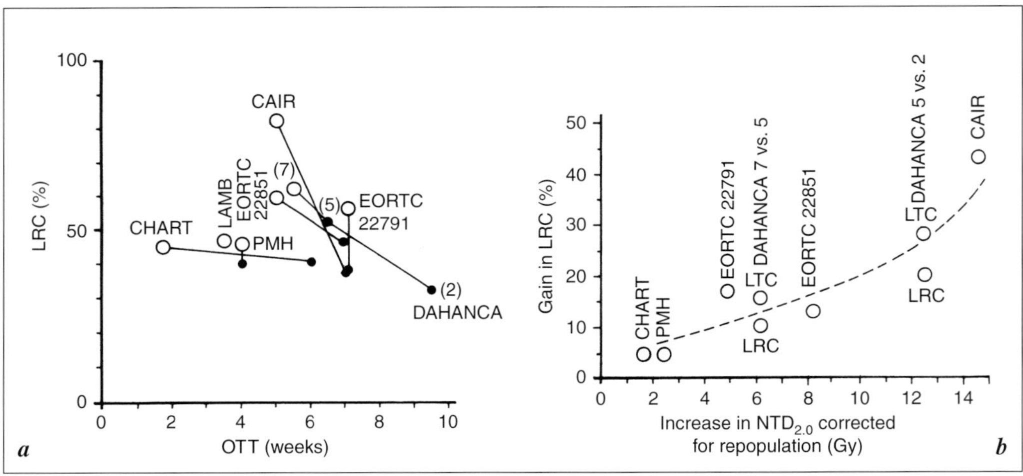


**Fig. 1.** LRC for different altered fractionation schedules depending on OTT (*a*) and normalized total dose if given in 2.0-Gy fractions ($NTD_2$) (*b*). The dose increment ($\Delta NTD_2$) was calculated in relation to NTD in the control arm.

is counteracted by accelerated repopulation. Thus, a therapeutic gain can be related either to dose intensity or to change in the OTT, or to both. It seems that the treatment benefit correlates with the magnitude of the shortening of the OTT but probably only beyond the 4th week of the treatment (fig. 1a). The increase in total physical dose (TD) seems poorly correlated with the gain in LRC, as it is influenced by other fractionation parameters. However, when the difference between TD given in the experimental arm and that of the conventional one is normalized to 2.0 fractions and corrected for repopulation rate (on average 0.6 Gy/day), then an extra 'effective' $NTD_2$ strongly correlates with therapeutic gain (fig. 1b).

Generally, the results of the most important clinical trials suggest that a therapeutic gain, at least for head and neck cancers, may be achieved by delivering as high total doses in short OTT as are tolerated by acutely and late-responding tissues. Furthermore, it also seems that there is no reason to shorten OTT to less than 4 weeks because it is unlikely to deliver higher total doses in very short courses without severely impairing acute and late tolerance.

## What Is the Price to Pay – Acute Toxicity?

The majority of trials on AF/HF documented an increased incidence and severity of acute mucosal reactions. Is it evidence-based that severe acute mucositis is dose-limiting for altered RT or is it only an assumption?

The incidence of confluent mucositis (CM) presented in table 1 is part of a large series carefully reviewed by Kaanders et al. [6]. From this review and table 1 some uncertainties arise. CM as the most severe morphological acute reaction still remains an undefined term. Scoring and counting of CM are likely based on a variety of interobserver differences, as well as on various objective and subjective aspects of the scoring systems. Incidence, onset and duration of a peak reaction, time to healing (which very likely reflects the degree of mucosal denudation), and functional disorders are those items which determine the overall severity of the acute mucosal reaction, but usually they are scored separately and with insufficient precision.

Despite these uncertainties, and except in a few studies (table 1, a–c, i) where acute effects were extremely severe, incidence and severity of CM with altered fractionation schedules were usually higher than in the conventional arm (table 1, d–h, k–l), but mostly acceptable even if the duration and time to complete healing were prolonged. It seems important whether the incidences correlate with the severity of CM, and whether they are both dose-dependent. It is well known that the CM incidence depends on the intensity of dose accumulation (accumulated dose, AD), at least as regards the total dose which is delivered during the first 2 weeks of treatment. Because during the 1st week of RT the discrepancy between cell production and cell killing is greatest, the choice of AD in the first week ($AD_1$) seems logical. However, for many altered regimes, except HF and short AF, the $AD_1$ varies quite widely (table 1). The $AD_1$ of 24 Gy in the EORTC 22851 [5] or even 31.5 Gy in the CHART [2] did not result in a particularly high incidence of CM. This may be explained by the fact that after 28.8 Gy in the EORTC schedule a 2-week break was introduced, whereas CHART was completed in 12 days. Although Peracchia and Salti [12] used a schedule very similar to CHART with $AD_1$ of 30 Gy, the incidence of CM was 100%, which resulted in 55% of consequential late effects (CLE). In this case, the high $AD_1$ was not the only reason, but mainly too short intervals between the daily fractions.

An $AD_1$ of 14–14.5 Gy (table 1, d, i, l) represents different fractionation schedules with a wide range of CM rate. In the escalated schedule of Harari [3], the $AD_1$ had the lowest value, which constantly increased in the consecutive weeks of treatment, reaching the highest value in the last week. The same situation characterizes concomitant AF, e.g. in the schedule of Kaanders et al. [6], where the $AD_1$ of 10 Gy (as for the conventional 2.0 Gy regimen) was associated with an unexpectedly high rate of CM. In the CAIR protocol, the $AD_1$ of 14 Gy accumulated during 7 days, including the weekend, resulting in an unacceptable rate of CLE. When the fraction size was lowered from 2.0 to 1.8 Gy ($AD_1 = 12.6$ Gy), no more CLE was observed.

All these results suggest that $AD_1$ is not a single good parameter predicting incidence and severity of acute effects. It seems also important how fast and at

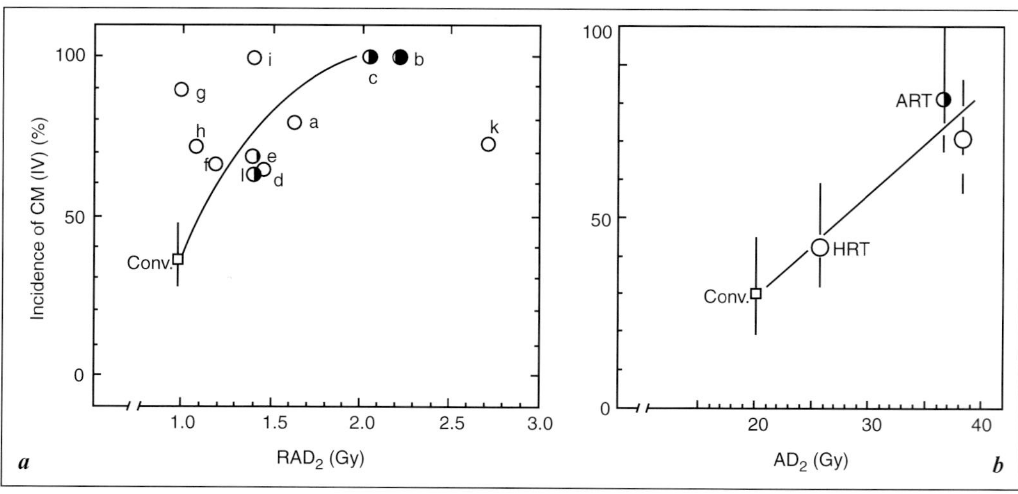



**Fig. 2.** Incidence of severe acute CM (IV). *a* Related to RAD in the 2nd week of treatment (RAD$_2$). RAD$_2$ was calculated as the ratio of given AD$_2$ applied to the AD$_2$ for conventional 2.0-Gy fractionation (AD$_{2conv.}$ = 20 Gy). The black area in the circles correlates with the rate of CLE (see table 1). *b* Average incidence of CM (IV) depending on type of altered fractionation. (The black area indicates the risk of CLE.)

what rate the dose is accumulated during the treatment. An AD$_1$ of 30 Gy could be acceptable for short HAF schedules provided that the time interval between fractions is at least 6 h, and that doses per fraction are lower than 2.0 Gy. An AD$_1$ of 20–26 Gy may represent any AF/HF schedule if it has a 2–3 weeks' break after the first 1–1.5 weeks of treatment, whereas for 6- and 7-day treatments AD$_1$ should not be higher than 12 Gy. In contrast, there is no single AD$_1$ value, which may characterize any concomitant boost AF schedule.

For the further analysis of the CM-AD relationship, the AD values for consecutive weeks of treatment (AD$_2$, AD$_3$ etc.) were used. Moreover, the AD was recalculated relative to the AD for conventional 2.0-Gy fractionation (RAD[2]). The results showed that the regression curve for RAD$_2$ (the end of the 2nd week) gives the best fit to the data analyzed (fig. 2a). When RAD$_2$ values were normalized for different $\alpha/\beta$ values in the range of 2–15 Gy, those for $\alpha/\beta$ of 2.0 Gy were the best representatives for the data sets. Although it is impossible to estimate accurate $\alpha/\beta$ values for CM based on available clinical data, the present results may indirectly suggest that mucosal tissues in the head and neck region may be more sensitive to change in dose per fraction than is generally

[2]RAD = AD$_n$/AD$_{n2.0}$ where n is a given week of treatment.

accepted. This suggestion may support the fact that in CAIR, a small (10%) change in fraction size from 2.0 to 1.8 Gy decreased the risk of CLE to zero.

Finally, both the present results and the review of Kaanders et al. [7] suggest that the 'dose-CM' relationship is complex in nature, and that there is no single parameter or factor, which could accurately predict a dose-limiting risk of CM. This prediction must currently still be mainly based on clinical experience and intuition. Except for pure intensive AF schedules using 2.0 fractions and insufficient time intervals between fractions ($\leq$4 h), in any other AF-HF schedule severity and incidence of CM, although higher than in conventional radiotherapy, does not appear to be dose-limiting (fig. 2b).

## How Optimal Could the Game Be Played?

Analyzing a large body of clinical data Withers et al. [13] have calculated that on average 0.6 Gy/day ($D_{rep}$) is lost due to accelerated tumor clonogen repopulation after a lag time of 28 days. However, there are some suggestions that the lag period might be shortened to 21 days and $D_{rep}$ may increase up to 1.0 Gy/day. On the other hand, mucosal epithelial cells begin to repopulate rapidly after 12 days, compensating on average 1.0 Gy/day during treatment, and even more (1.8–2.5 Gy) during treatment-free intervals (breaks, weekends). Generally, these average values for tumors and acute effects were calculated for 5-day/week treatments and all alterations in fractionation schedules were also designed within the period of 5 working days, i.e. excluding weekends.

The first step to extend the treatment to weekends was made in the DAHANCA-7 trial (6-day schedule) which resulted in a 10% benefit in LRC (16% in LTC), due to shortening the OTT from 6.5 to 5.5 weeks. The CAIR trial [9] included the entire weekend for treatments, thus accelerating the treatment from 7 to 5 weeks by giving one fraction per day with constant 24-hour intervals. It yielded a 45% increase in 3-year LRC, which gives a 22.5% benefit for 1-week shortening of the OTT. This shows that a 10% benefit of the Saturday-DAHANCA schedule is doubled by the Saturday-Sunday CAIR schedule. It also indirectly suggests that $D_{rep}$ could be higher on treatment-free weekends compared to treatment days during the week. Therefore, it might likely be assumed that between Friday and Monday accelerated repopulation compensates for 0.9 Gy/day compared to only 0.4 Gy/day between Monday and Friday (on average 0.6 Gy/day). According to this hypothetical assumption, the weekly dose balanced by repopulation would depend on the number of treatment days and duration of the interfraction interval; and thus, for a 3-day/week treatment [8] it would be 5.3 Gy, 4.2 Gy for a 5-day/week treatment, 3.5–3.85 Gy for a 6-day/week schedule (sixth fraction is given on Friday evening or on Saturday morning)

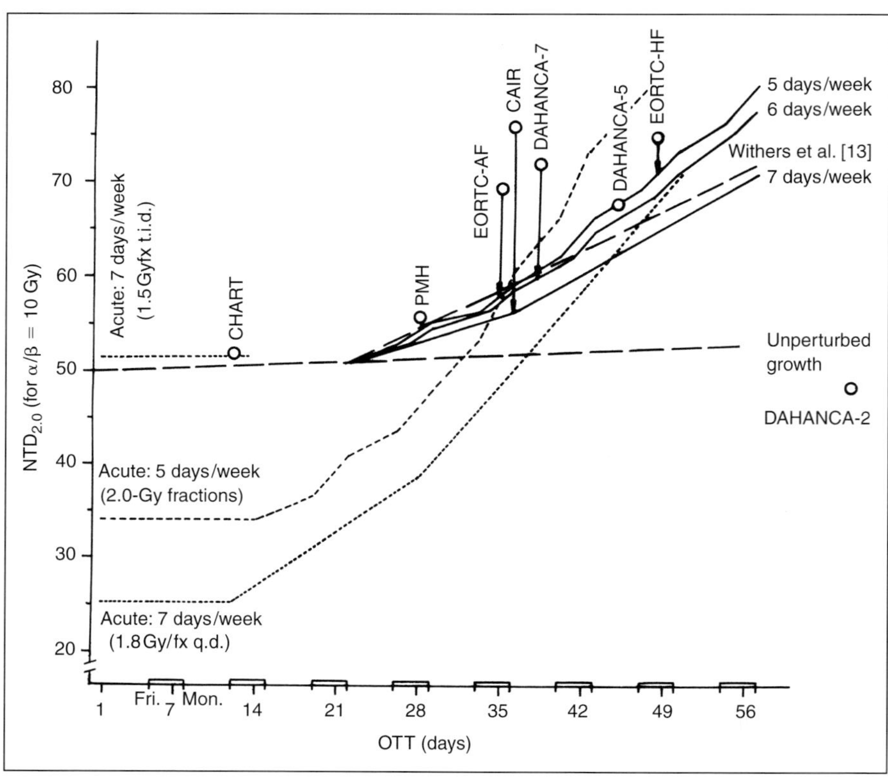


***Fig. 3.*** Theoretical modeling of isoeffect curves for head and neck cancers and acutely responding normal tissues depending on different treatment intensity (5–7 days/week), and a different repopulation rate during working days and weekends based on the following assumption. Tumor: TCD$_{50}$ equals 50 Gy in the absence of repopulation. After 21 days it increases by 0.4 Gy/day during Monday to Friday and 0.9 Gy/day during Saturday to Monday until week 5, and 0.7 Gy/day and 1.4 Gy, respectively, thereafter. Acute effects: The tolerance dose equals 34 Gy in 2-Gy fractions (5 days/week) in the absence of repopulation [13], and after 14 days it increases by 0.7 Gy/day during Monday to Friday and 1.2 Gy/day during Saturday to Monday until week 4, and 1.4 and 2.4 Gy/day thereafter. For a 7-day treatment, tolerance dose equals 25.2 Gy given in 1.8 Gy/fx q.d. during the first 12–14 days or it is 51 Gy if given in 1.5 Gy/fx in 12 days as a whole course of treatment. For 8 different altered schedules, the normalized total dose (NTD$_{2.0}$) values corrected for dose/fraction and OTT are related to the respective isoeffect curves.

and only 2.8 Gy for a 7-day/week schedule. The respective theoretical dose-time curves, shown in figure 3, give a better fit to the NTD values of the selected trials and better correlate with the respective LRC rates than an average curve proposed by Withers et al. [13]. Thus, an optimal therapeutic benefit might be expected for schedules with the OTT of 4–5.5 week, and there is no reason to shorten or to extend the OTT below or above that range. Similarly, normal mucosa should show this tendency, but with an earlier onset, and a higher and faster rate of repopulation, and thus the higher risk and severity of acute mucosal reactions with an increasing number of treatment days per week should not be surprising.

Doses around 60 Gy in 4 weeks or even 70 Gy in 5 weeks seem to be within the acceptable risk of severe acute reactions [7, 13]. Any breaks, including weekends, especially in the second part of treatment, immediately decrease the intensity of accumulation of effective dose and hence reduce the expected benefit. Provided that intervals between fractions are sufficiently long ($\geq$6 h) and fraction sizes respectively low, such treatments can only be delivered if they are extended beyond working hours and days. However, it should also be remembered that even small variations in total dose, dose per fraction and interfraction intervals can enhance originally tolerable acute effects into very severe acute and into intolerable CLE.

Although there are some reasonable suggestions on how the optimal game should be played, the choice of the most effective altered fractionation schedule for specific sites and stages of head and neck cancers still remains an open question.

## References


1   Cummings BJ, O'Sullivan B, Keane T, Gullane P: Larynx conservation in a randomized trial hyperfractionated (HFRT) versus conventional once daily radiation (CRT): A subgroup analysis. Eur J Cancer 1997;33(suppl 8):254.
2   Dische S, Saunders MC, Barrett A, Harvey A, Gibson D, Parmer M: A randomized multicentre trial of CHART versus conventional radiotherapy in head and neck cancer. Radiother Oncol 1997; 44:123–136.
3   Harari PM: Adding dose escalation to accelerated hyperfractionation for head and neck cancer: 76 Gy in 5 weeks. Semin Radiat Oncol 1992;2:58–61.
4   Horiot JC, LeFur R, Nguyen A, et al: Hyperfractionation versus conventional fractionation in oropharyngeal carcinoma. Final analysis of a randomized trial of the EORTC cooperative group in radiotherapy. Radiother Oncol 1992;25:231–241.
5   Horiot JC, LeFur R, Schraub S, et al: Status of the experience of the EORTC Cooperative Group of Radiotherapy with hyperfractionated and accelerated radiotherapy regimes. Semin Radiat Oncol 1992;2:34–37.
6   Kaanders JHAM, van Daal WAJ, Hoogenraad WJ, van der Kogel AJ: Accelerated fractionation for laryngeal cancer, acute and late toxicity. Int J Radiat Oncol Biol Phys 1992;24:497–503.
7   Kaanders JHAM, van der Kogel AJ, Ang KK: Altered fractionation: Limited by mucosal reactions. Radiother Oncol 1998;50:247–260.
8   Lamb D, Spry N, Gray A, et al: Accelerated fractionation radiotherapy for advanced head and neck cancer. Radiother Oncol 1990;18:107–116.
9   Maciejewski B, Składowski K, Tarnawski R, Zajusz A: Is the 7-day accelerated treatment better than conventional 5-day irradiation of head and neck cancers?; in Kogelnik HD, Sadlmayer F (eds): Progress in Radio-Oncology. Monduzzi, 1998, vol 6, pp 767–773.
10  Nguyen T, Demange L, Froissart D, et al: Rapid hyperfractionated radiotherapy. Clinical results in 178 advanced squamous cell carcinomas of the head and neck. Cancer 1985;56:16–19.
11  Overgaard J, Saud H, Overgaard M, et al: Importance of overall treatment time for the outcome of radiotherapy in head and neck carcinoma. Experience from the Danish Head and Neck Cancer Study; in Kogelnik HD, Sadlmayer F (eds): Progress in Radio-Oncology. 1998, vol 6, pp 743–752.

12  Peracchia G, Salti C: Radiotherapy with thrice-a-day fractionation in a short overall time. Clinical experience. Int J Radiat Oncol Biol Phys 1981;7:99–104.
13  Withers HR, Maciejewski B, Taylor JMG: Biology of options in dose fraction; in McNally NJ (ed): BIR Report 19: The Scientific Basis of Modern Radiotherapy. London, British Institute of Radiology, 1989, pp 27–36.

Prof. B. Maciejewski, Cancer Center, MSC Institute,
Ar. Krajowej, 15, PL–44-101 Gliwice (Poland)
Tel. +48 32 231 54 87, Fax +48 32 230 78 07, E-Mail maciejewski@onkol.instonko.gliwice.pl

Dörr W, Engenhart-Cabillic R, Zimmermann JS (eds): Normal Tissue Reactions in Radiotherapy and Oncology. Front Radiat Ther Oncol. Basel, Karger, 2002, vol 37, pp 185–190

# Side Effects in Multimodal Treatment Concepts in Carcinomas of the Head and Neck

*B.M. Lippert, A.A. Dünne, J.A. Werner*

Department of Otolaryngology, Head and Neck Surgery,
Philipps University of Marburg, Germany

Detailed knowledge of possible side effects of multimodal treatment concepts for squamous cell carcinomas of the head and neck region is essential, because they determine the choice of the individual treatment modality. According to their localization, carcinomas of the upper aerodigestive tract influence vital functions such as breathing and nutrition. Moreover, the important communication function, speaking, must be considered. With regard to tumor-related functional impairment, such as dysphagia and odynophagia, the different side effects expected after the different multimodal treatment strategies are highly relevant. Moreover, comorbidity of the usually older patients, often complicated by long-lasting consumption of tobacco and alcohol, has to be considered. It is very likely that the individual treatment concept has a direct impact on the quality of life.

The treatment of malignant tumors of the upper aerodigestive tract can include surgery, radiotherapy and chemotherapy. These treatment modalities are applied alone or, in most cases of squamous cell carcinomas, as combination therapy, depending on tumor-related factors, such as localization, size and existing metastases, and on the intention of therapy, i.e. a curative or palliative approach.

The most common tumors of the mucous membranes of the upper aerodigestive tract are squamous cell carcinomas; this summary focuses exclusively on them. The severity of side effects of multimodal treatment modalities for squamous cell carcinomas is influenced by several factors, including radiation dose, chemotherapeutic agents and their dose, but also individual radiosensitivity, general health status and nutritional status of the patient.

## Surgical Treatment

Surgical treatment of squamous cell carcinomas of the head and neck includes resection of the primary tumor and removal of the locoregional lymph nodes (neck dissection). The side effects of surgical therapy vary depending on the localization and the extent of the primary tumor. Over the last 2 decades, a considerable change in treatment policy has been observed due to the introduction of laser therapy [6]. According to oncological criteria, organ-preserving resections can frequently be performed, resulting in a significant improvement of the postoperative quality of life.

The surgical treatment of regional lymph node metastases is based on early work by Crile [3] who in 1906 recommended radical neck dissection based on the results of 132 surgeries. This approach includes removal of all cervical lymph nodes located in regions I–V as well as removal of the accessory nerve, the internal jugular vein and the sternocleidomastoid muscle. This extended surgery is associated with significant side effects with considerable functional impairment. Limitation of the neck dissection aiming at minimization of the side effects, depicted as functional neck dissection, was first described by Suárez [7] in 1963. This type of neck dissection should preserve a maximum of functions without affecting prognosis [8]. Today, the principle of selective neck dissection is surgical removal of single lymph node regions depending on the localization of the primary tumor.

*Side Effects of Neck Dissection*
The risk of postoperative side effects associated with the removal of cervical lymph nodes directly correlates with the extent of the chosen type of neck dissection. This includes wound infection and delayed wound healing, the injury of vessels or nerves, the development of chyle fistula or chylothorax, of lymphedema or of a shoulder-hand syndrome, fractures of the clavicle, formation of cicatrical pulls, keloids and contractures. Ligation of the internal jugular vein may increase the intracranial pressure, particularly after bilateral radical neck dissection. Clinical symptoms can be headache, nausea and vomiting. Much less frequently, loss of vision or blindness are seen [9].

*Wound Infection.* Wound infection and delayed wound healing (fig. 1) are the most frequent complications after neck dissection. Despite perioperative administration of antibiotics, their incidence is 10–20%. Preoperative radio- or radiochemotherapy significantly increases the risk of postoperative wound infections. The impairment of wound healing can result in functional morbidity, prolonged hospitalization, or a delay of postoperative radiotherapy with a higher risk of tumor recurrence. Prolonged operation time and large wound defects,

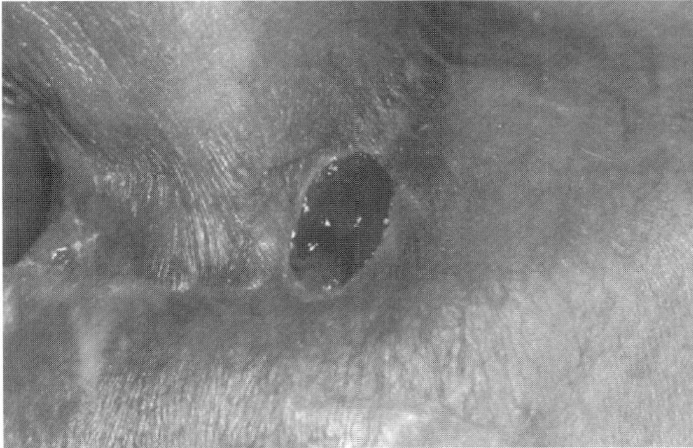

***Fig. 1.*** Significant wound healing impairment with fistulation in the area of the right side of the neck as a consequence of laryngectomy and bilateral modified radical neck dissection plus radiochemotherapy.

especially in reconstructive surgery with microanastomosed distant flaps or intestinal grafts, increase the risk of subsequent wound healing impairment.

*Vessel Injury.* Traumatization of large venous vessels can lead to life-threatening bleeding and air embolism. Thrombosis of the internal jugular vein after mechanic trauma at the surface or intraoperative desiccation, as well as ligature of the internal jugular vein and/or involvement of lymphatic vessels consequently can lead to a lymphatic edema. Maximum edema is usually observed after 3–4 days followed by regression after 7–10 days. Destruction of anchor filaments of subepithelial lymphatic vessels promotes edema formation. With multimodal therapy consisting of surgery and postoperative radiochemotherapy, additional effects resulting in prolongation and intensification of the initial edematous response (fig. 2) must be considered. Because of extended laryngeal and pharyngeal edemas, prophylactic tracheotomy must be discussed in single cases according to the aggressiveness of the therapy regime (type of neck dissection, radiation dose, chemotherapy).

*Injury of Neural Structures.* The injury of neural structures is a frequent complication after a neck dissection. In radical neck dissection the accessory nerve is removed and >60% of patients develop a shoulder-hand syndrome, which is characterized by the incapability to lift the arm to more than 90°. Removal of neck level I can affect the submandibular branches of the facial nerve. An unilateral lesion of the phrenic nerve, which is observed in about 8% of patients, is clinically irrelevant. Bilateral damage, however, can result in

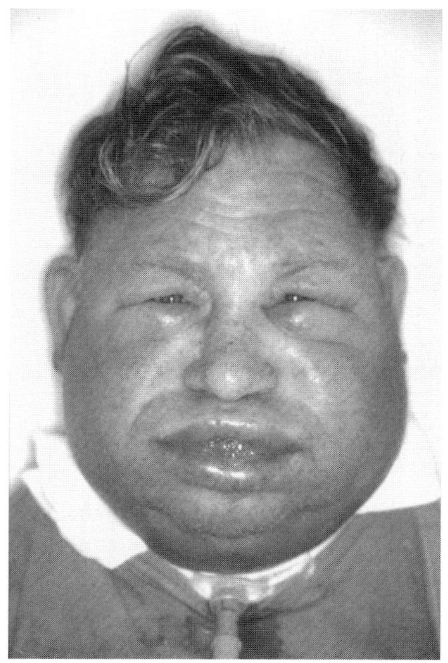

***Fig. 2.*** Submental and submandibular lymphedema after laryngectomy, bilateral selective neck dissection and postoperative radiochemotherapy.

significant respiratory symptoms due to diaphragmatic eventration. Further complications are damage of the hypoglottic nerve of both sides, leading to aphagia. In these cases a percutaneous gastrostomy must be performed in order to guarantee proper nutrition. Percutaneous gastrostomy itself can be associated with side effects like peritonitis, wound infection and perforation. Damage of the sympathetic trunk leads to Horner's syndrome and damage of the brachial plexus results in reduced mobility of arm and fingers.

*Functional Disorders.* Particularly after a radical neck dissection, cicatricial pulls and contractures are observed. These can be increased significantly by postoperative radiochemotherapy. Devascularization of the clavicle due to surgery or irradiation can result in fractures and aseptic bone necrosis. In the last years the limitation of the extent of neck dissection resulted in a significant reduction of side effects, while comparable 5-year survival rates are observed after modified radical neck dissection and selective neck dissection [4]. Moreover, after selective neck dissection in comparison to modified radical neck dissection the function of the shoulder is better and fewer thromboses of the internal jugular vein occur [2, 5]. In conclusion, a significantly lower rate of postoperative pain, shoulder dysfunction and torsion of the shoulder blade is observed in addition to a better quality of life after modified radical and selective neck dissection in comparison to a radical neck dissection [1].

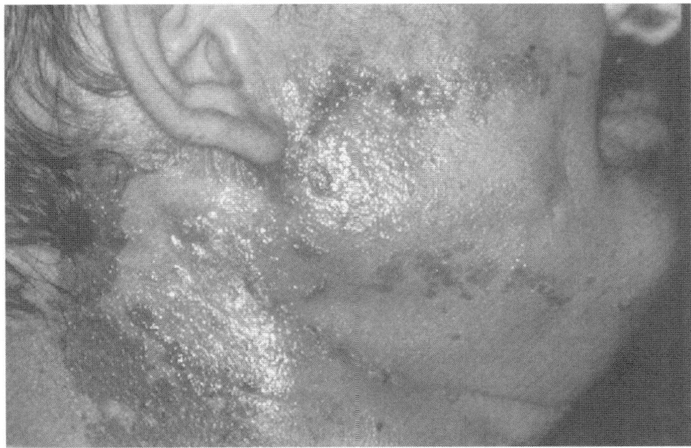

***Fig. 3.*** Radiodermatitis. Epitheliolysis during radiochemotherapy (day 17 of radio-therapy).

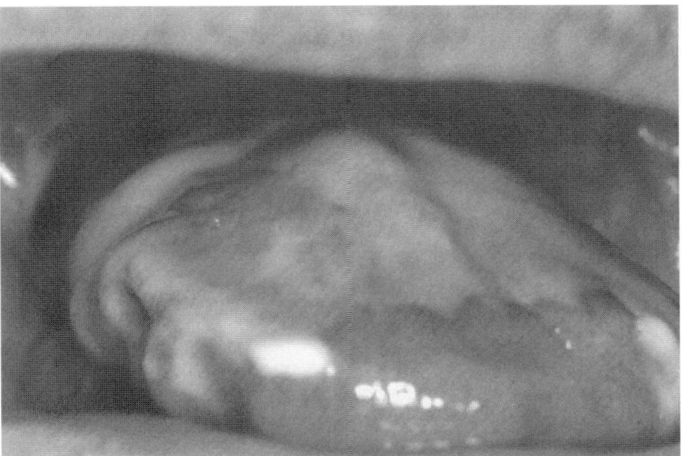

***Fig. 4.*** Confluent oral mucositis during radiochemotherapy.

## Radiotherapy

Independent of the complications of surgical treatment, radiation dose and combined chemotherapy correlate with the extent of the side effects in skin and mucosal membranes. These manifest as radiodermatitis (fig. 3) or mucositis (fig. 4).

## Multimodal Treatment

Potential side effects of multimodal treatment modalities of carcinomas of the upper aerodigestive tract have a direct influence on the quality of life and the way of living of the patient. The choice of the individual treatment concept must be based on the general situation of the patient and his prognosis with a close cooperation between surgeons and radiologists.

The introduction of organ-preserving treatment concepts and the limitation of neck dissection reduced surgery-related side effects in the multimodal treatment concept of carcinomas of the head and neck without impact on tumor effects. Prospective studies are now necessary to establish the selective type of neck dissection in the surgical treatment of squamous cell carcinomas of the head and neck.

## References


1   End results of a prospective trial on elective lateral neck dissection vs. type III modified radical neck dissection in the management of supraglottic and transglottic carcinomas. Brazilian Head and Neck Cancer Study Group. Head Neck 1999;21:694–702.
2   Cheng PT, Hao SP, Lin YH, Yeh AR: Objective comparison of shoulder dysfunction after three neck dissection techniques. Ann Otol Rhinol Laryngol 2000;109:761–766.
3   Crile GW: Excision of cancer of the head and neck with a special reference to the plan of dissection based upon one hundred thirty-two operations. JAMA 1906;47:1780–1786.
4   Kuntz AL, Weymuller EA Jr: Impact of neck dissection on quality of life. Laryngoscope 1999;109:1334–1338.
5   Prim MP, De Diego JI, Fernandez-Zubillaga A, Garcia-Raya P, Madero R, Gavilan J: Patency and flow of the internal jugular vein after functional neck dissection. Laryngoscope 2000;110:47–50.
6   Steiner W: Therapie des Hypopharynxkarzinoms. 1. Literaturübersicht Chirurgie und/oder Radiotherapie. HNO 1994;42:4–13.
7   Suárez O: El problema de las metástasis linfáticas y alejadas del cáncer de laringe e hipofaringe. Rev Otorhinolaringol 1963;23:83–99.
8   Werner JA: Historischer Abriss zur Nomenklatur der Halslymphknoten als Grundlage für die Klassifikation der Neckdissektion. Laryngorhinootologie 2001;80:400–409.
9   Werner JA: Aktueller Stand der Versorgung des Lymphabflusses maligner Kopf-Hals-Tumoren. Eur Arch Otorhinolaryngol 1997(suppl 1):47–85.



Burkard M. Lippert, MD, Department of Otolaryngology, Head and Neck Surgery,
Philipps University Marburg, Deutschhausstrasse 3, D–35037 Marburg (Germany)
Tel. +49 6421 2862835, Fax +49 6421 2866367, E-Mail lippert@mailer.uni-marburg.de

Dörr W, Engenhart-Cabillic R, Zimmermann JS (eds): Normal Tissue Reactions in Radiotherapy
and Oncology. Front Radiat Ther Oncol. Basel, Karger, 2002, vol 37, pp 191–195

# Clinical Side Effects after Radical Prostatectomy

*Rolf von Knobloch, Sebastian Wille, Rainer Hofmann*

Department of Urology, Philipps University Medical School, Marburg, Germany

In western European countries prostate cancer has become the leading cause of cancer death. With no effective systemic treatment of prostate cancer available, the best chance of decreasing mortality is to provide curative treatment while the tumor is still localized. Treatment choices for organ-confined disease include either radiotherapy or radical prostatectomy. In patients with a cancer-independent life expectancy of at least 10 years and under the age of 75, radical prostatectomy is the primary treatment option. The rate of radical prostatectomy has increased 6-fold between 1984 and 1990 due to aging of the population, prostate-specific antigen-based prostate cancer screening, development of transrectal ultrasound making it easier to perform biopsies, and major improvements in surgical techniques [2]. Despite the detection of cancer at earlier stages and the more frequent application of radical prostatectomy, mortality rates from prostate cancer have not changed significantly. Until there are markers available to reliably predict the biological behavior of prostate cancers, the primary choice of treatment in patients with localized disease and a reasonable life expectancy is curative radical prostatectomy or radiotherapy rather than watchful waiting. Even though radical prostatectomy can lead to side effects, such as urinary incontinence and erectile dysfunction, worse than early cancer-related symptoms, the surgical procedure still offers best local tumor control and, due to the histopathology obtained, has a high predictive value for the further outcome. In order to assess the risk of the specific side effects and morbidity from radical prostatectomy and their influence on quality of life we retrospectively analyzed patients treated with radical retropubic prostatectomy (RRP) for prostate cancer at our institution between 1989 and 1996 and additionally included results of a large contemporary RRP series from the literature.

## Materials and Methods

The charts of patients treated at our institution with RRP for cancer of the prostate from 1989 to 1996 were reviewed for operative and perioperative complications and for treatment-related specific side effects, namely urinary stress incontinence, impotence and anastomotic stricture. Results of recent investigations on side effects of RRP were included to make large scale statements on the risk of these side effects. The influence of RRP-related side effects on the quality of life is discussed. To obtain comparable results of different investigations on the issue of postoperative continence we only evaluated groups of patients with total continence and complete urinary stress incontinence (grade III) where possible. The rate of postoperative impotence was evaluated in relation to the number of neurovascular bundles preserved intraoperatively and patient age at the time of surgery.


## Results

In Marburg 449 patients between 1989 and 1996 received RRP for prostate cancer. The follow-up for side effects ranged from 1 to 65 months, with a median follow-up of 9.25 months. Perioperative complications were rare with an incidence of 5.6%. Main causes of perioperative complications requiring treatment were of cardiopulmonary and thromboembolic origin with an incidence of 2.4% each. Treatment-specific complications, i.e. anastomotic insufficiency merely requiring prolonged catheter drainage for 21 rather than 7 days, symptomatic lymphocele and rectal lesions, occurred in less than 4.8, 1.7 and 0.9%, respectively. There was no case of perioperative mortality. In the literature there are comparable perioperative complication rates of 10% and an overall mortality rate of 0–1% [1, 2, 5]. A Johns Hopkins series from 1994 reports an average blood loss of 1,490–1,940 ml during RRP, requiring transfusions of 730–960 ml of blood [8].

Treatment-related side effects of RRP are reported in a varying incidence. In the Marburg series with a short mean follow-up of 9.25 months (range 1–65 months) patients complained about stress urinary incontinence grades II and III in 15%. On discharge from hospital directly after primary treatment, the rate of stress urinary incontinence of grade I or higher was as high as 73%. In the literature complete incontinence is reported with incidences of 0.6–34.8% [1, 2, 4–7, 10] with complete continence rates of 75–92% (table 1). The time of follow-up is the crucial issue when evaluating postoperative incontinence as it takes up to 18 months for the patients to regain continence. Furthermore, younger patients show lower rates of postoperative incontinence [1, 2, 4, 5, 7]. Anastomotic strictures requiring intervention occur in 6–20.5% of patients after RRP (table 2). High rates of anastomotic strictures are observed in series with extensive operative reconstruction of the bladder neck in favor of low incidences of postoperative incontinence [4–6, 10].

*Table 1.* RRP-related side effects: postoperative urinary stress incontinence

| Reference | Patient number[a] | Independent survey[b] | Follow-up months | Complete continence % | Total incontinence % |
|---|---|---|---|---|---|
| Marburg | 449 | no | 1–65 | 45 | 15 ($>$I)[c] |
| Catalona et al. [1] | 1,325 | no | $>$18 | 92 | 8 |
| de Kernion et al. [2] | review | no | $>$3 | 75 | 0.6 |
| Geary et al. [4] | 458 | no | $>$12 | 80.1 | 5.2 |
| Huland [5] | review | no | n.i. | 80–90 | 4 |
| Kao et al. [6] | 1,069 | yes | $>$6 | 33 | 10.4 ($>$I)[c] |
| Richie [7] | review | no | n.i. | 92 | 2 |
| Stanford et al. [10] | 1,291 | yes | 24 | n.i. | 8 |

n.i. = No information.
[a] Number of patients analyzed in study.
[b] Independent survey by means of uncommented questionnaire.
[c] All patients with incontinence $>$ grade I included.

*Table 2.* RRP-related side effects: postoperative stricture of the vesicourethral anastomosis

| Reference | Patient number[a] | Independent survey[b] | Follow-up months | Anastomotic stricture % |
|---|---|---|---|---|
| Catalona et al. [1] | 1,325 | no | $>$18 | 4 |
| de Kernion et al. [2] | review | no | $>$3 | 1 |
| Geary et al. [4] | 458 | no | $>$12 | 17.5 |
| Huland [5] | review | no | n.i. | $<$6 |
| Kao et al. [6] | 1,069 | yes | $>$6 | 20.1 |
| Stanford et al. [10] | 1,291 | yes | 24 | 16.1 |

n.i. = No information.
[a] Number of patients analyzed in study.
[b] Independent survey by means of uncommented questionnaire.

A further side effect of RRP causing patient discomfort is impotence. Erectile dysfunction is hard to assess and depends on factors such as preoperative potency, age of the patient at surgery, time of follow-up after surgery and number of preserved neurovascular bundles. In our series this issue was not investigated due to inconsistent information of pre- and postoperative potency. Results reported for postoperative impotence in the literature [1, 2, 4–7, 10] vary between 16 and 77% depending mainly on the application of nerve-sparing

| Reference | Patient number[a] | Independent survey[b] | Follow-up months | Impotence % |
|---|---|---|---|---|
| Catalona et al. [1] | 1,325 | no | >18 | 32–53[c] |
| de Kernion et al. [2] | review | no | >3 | 20–45[c] |
| Richie [7] | review | no | n.i. | 24–40[c] |
| Kao et al. [6] | 1,069 | yes | >6 | 77[d] |
| Stanford et al. [10] | 1,291 | yes | 24 | 56–66[e] |

n.i. = No information.
[a] Number of patients analyzed in study.
[b] Independent survey by means of uncommented questionnaire.
[c] Preservation of two or one neurovascular bundle.
[d] No data on nerve sparing.
[e] Nerve-sparing and no nerve-sparing technique.

techniques with the preservation of one or both neurovascular bundles as well as the average age of the patients at the time of surgery (table 3).

## Discussion

Reported incidences of RRP-related side effects vary to a great extent. Reasons for the different rate of side effects are the surgeon's and/or center's experience, modifications of the surgical procedure, for instance the adoption of nerve-sparing techniques to preserve potency, definiton of side effects, especially for urinary incontinence, time of follow-up after RRP, and the person conducting the inquiry into complications, i.e. surgeons and assistants themselves or independent investigators. Until prospective randomized multicenter studies compare radical prostatectomy with radiotherapy in the treatment for localized prostate cancer providing data on treatment-related side effects, the quoted side effects may continue to vary considerably. Nevertheless, analysis of the influence of RRP-related side effects on patients' quality of life shows that 81% of patients were satisfied with the surgical outcome and a total of up to 89% of all patients would again choose surgery as therapy option [3, 9, 10]. In a study comparing different treatment modalities for localized prostate cancer 6 years after diagnosis, the majority of men were bothered by an impairment of their sexual function regardless of treatment, while only a minority complained about their current urinary function [9].

In conclusion, RRP for localized prostate cancer in experienced hands offers the best options for local tumor control with acceptable rates of postoperative urinary stress incontinence, impotence and anastomotic stricture. The influence of RRP on patients' quality of life is moderate, with up to 89% of patients stating they would again choose surgery as their treatment.

## References


1   Catalona WJ, Carvalhal GF, Mager DE, Smith DS: Potency, continence and complication rates in 1,870 consecutive radical retropubic prostatectomies. J Urol 1999;162:433–438.
2   de Kernion J, Belldegrun A, Naitoh J: Surgical treatment of localized prostate cancer: Indications, technique and results; in Belldegrun A, Kirby RS, Oliver T (eds): New Perspectives in Prostate Cancer. Oxford, Isis Medical Media, 1998.
3   Fowler FJ Jr, Barry MJ, Lu-Yao G, Wasson J, Roman A, Wennberg J: Effect of radical prostatectomy for prostate cancer on patient quality of life: Results from a Medicare survey. Urology 1995;45:1007–1013.
4   Geary ES, Dendinger TE, Freiha FS, Stamey TA: Incontinence and vesical neck strictures following radical retropubic prostatectomy. Urology 1995;45:1000–1006.
5   Huland H: The risks outweigh the benefits of radical prostatectomy in localised prostate cancer: The argument against. Eur Urol 1996;29(suppl 2):31–33.
6   Kao TC, Cruess DF, Garner D, et al: Multicenter patient self-reporting questionnaire on impotence, incontinence and stricture after radical prostatectomy. J Urol 2000;163:858–864.
7   Richie JP: Localized prostate cancer: Overview of surgical management. Urology 1997;49(suppl 3A):35–37.
8   Shir Y, Raja SN, Frank SM, Brendler CB: Intraoperative blood loss during radical retropubic prostatectomy: Epidural versus general anesthesia. Urology 1995;45:993–999.
9   Smith DS, Carvalhal GF, Schneider BAK, Krygiel J, Yan Y, Catalona WJ: Quality-of-life outcomes for men with prostate carcinoma detected by screening. Cancer 2000;88:1454–1463.
10  Stanford JL, Feng Z, Hamilton AS, et al: Urinary and sexual function after radical prostatectomy for clinically localized prostate cancer: The Prostate Cancer Outcomes Study. JAMA 2000;283:354–360.



Dr. Rolf von Knobloch, Department of Urology, Philipps-Universität Marburg,
Baldingerstrasse, D–35033 Marburg (Germany)
Tel. +49 6421 286 2513, Fax +49 6421 286 5590,
E-Mail r.von-knobloch@mailer.uni-marburg.de

Dörr W, Engenhart-Cabillic R, Zimmermann JS (eds): Normal Tissue Reactions in Radiotherapy and Oncology. Front Radiat Ther Oncol. Basel, Karger, 2002, vol 37, pp 196–199

# Prostate Cancer – The Royal Marsden Conformal Experience

*Christopher M. Nutting, David P. Dearnaley*

Academic Unit of Radiotherapy and Oncology, Institute of Cancer Research and Royal Marsden NHS Trust, Sutton, UK

The aims of radical (curative) radiotherapy are to deliver a homogeneous radiation dose to a tumour target while minimising the dose to surrounding normal tissues. In this way the maximum number of tumour cells can be eradicated with the minimum risk of normal tissue injury. To achieve these aims, computed tomography (CT) planning with three-dimensional reconstruction of volumes of interest, clear definition of treatment margins, accurate patient positioning, and meticulous treatment, verification procedures are necessary. Shaped fields are produced by customised shaped blocks or multileaf collimation.

In the past, the tumour localisation was based on clinical examination or postoperative reports; alternatively a class of solutions was used for each tumour type where the borders of the radiation fields were defined in relation to standard anatomical landmarks. It has become apparent that in some cases these techniques of treatment planning did not encompass the tumour precisely within the high dose volume, or that the delivered dose to parts of the tumour was inhomogeneous. In addition, these conventional radiotherapy techniques used rectangular or simply shaped beams that for many tumour sites led to irradiation of unnecessarily large volumes of normal tissue.

## Rationale for Conformal Radiotherapy

Three-dimensional conformal radiotherapy (3DCRT) aims to overcome some of the problems associated with conventional radiotherapy. The tumour and surrounding normal tissue structures are visualised in three dimensions reconstructed from transaxial CT images of the patients' anatomy. This reduces the chance of missing extensions of the tumour which are not clinically detectable, and allows prediction of the radiation dose to surrounding radiosensitive organs. Each radiation beam is shaped such that normal tissues close to the tumour are shielded from radiation. For prostate cancer the amount of normal tissue treated to the 90% isodose may be reduced by 42%, with 46 and 41% reductions in the volumes of bowel and bladder, respectively [2]. The resultant reduction of the volume of normal tissue irradiated reduces the risk of radiation damage to these organs, which is known as the volume effect. If equivalent radiation doses are given, then a reduction in side effects is seen. Alternatively, for the same complication rate, an escalation of the delivered dose to a tumour may be possible, enhancing the therapeutic ratio. The paper illustrates several clinical studies carried out at the Institute of Cancer Research and Royal Marsden NHS Trust on patients with prostate cancer, and discusses future directions.

## A Randomised Trial of Conventional and Conformal Radiotherapy [1]

This trial was designed to test the hypothesis that conformal radiotherapy reduces the late side effects of irradiation because it allows a smaller amount of rectum and bladder to be exposed, by shaping the high-dose volume to the prostate.

Two hundred and twenty-five men with prostate cancer (stage T1–4, G1–3, N0, M0) were treated with radiotherapy to a standard dose of 64 Gy in daily 2-Gy fractions. The men were randomly assigned to receive conformal or conventional radiotherapy. The primary endpoint was the development of late radiation complications (>3 months after treatment) measured with the Radiation Therapy and Oncology Group (RTOG) score. Indicators of tumour control were also recorded.

Significantly fewer men developed radiation-induced proctitis and bleeding in the conformal group than in the conventional group (37 vs. 56% ≥RTOG grade 1, p = 0.004; 5 vs. 15% ≥RTOG grade 2, p = 0.01). There were no differences between groups in bladder function after treatment (53 vs. 59% ≥RTOG grade 1, p = 0.34; 20 vs. 23% ≥RTOG grade 2, p = 0.61). After median follow-up of 3–6 years there was no significant difference between groups in local tumour control [conformal 78% (95% CI 66–86), conventional 83% (69–90)]. The interpretation of this study was that conformal techniques significantly lowered the risk of late radiation-induced proctitis after radiotherapy for prostate cancer.

## A Randomised Trial of Dose Escalation in Prostate Cancer

Our current randomised study tests the hypothesis that higher doses of radiation can increase local control rates of prostate cancer, with the potential to improve overall survival. By using conformal techniques, the late side effect profile should be minimised. The trial design is a randomised study comparing 74 Gy with 64 Gy in conjunction with neoadjuvant androgen deprivation [5]. No validated data are available from this trial yet.


## Patient Immobilisation

Patient immobilisation is critical for accurate field placement in conformal radiotherapy. We performed a randomised study [3] in 30 men receiving radical radiotherapy for prostate cancer to evaluate a customised immobilisation system (IMS) on field placement accuracy (measured using an electronic portal imaging device), treatment delivery time, radiographer and patient acceptability. Patients were randomised using a crossover trial design to have radiotherapy planning and treatment given either in a conventional treatment position (CTP) or using an IMS.

Median simulation and treatment times were 22.5 and 9 min in the CTP and 25 and 10 min for the IMS ($p < 0.001$ for both). Median isocenter displacement for anterior fields was 1.7 mm from the simulated isocenter for the CTP compared to 2.0 mm for IMS ($p = 0.07$). For lateral and anterior fields no clinically significant reduction in either systematic or random field placement errors was demonstrated. The IMS was more comfortable than CTP ($p < 0.001$), but caused greater difficulty in patient positioning ($p < 0.001$), and alignment to skin tattoos ($p < 0.001$). We concluded that although IMS may have been more comfortable, treatment accuracy was not improved compared to the CTP in our department, and we currently treat patients with ankle stocks, but no pelvic immobilisation.


## Intensity-Modulated Radiotherapy

We are currently evaluating the application of intensity-modulated radiotherapy (IMRT) in prostate cancer. We have investigated the potential of IMRT to irradiate the prostate gland and pelvic lymph nodes while sparing critical pelvic organs [4]. Optimised conventional radiotherapy and 3DCRT plans were created and compared to inverse-planned IMRT. With conventional radiotherapy the mean percentage volume of small bowel and colon receiving >45 Gy was 21%. For 3DCRT it was 18% ($p = 0.0043$) and for 9-field IMRT it was 5% ($p < 0.001$ compared to 3DCRT). When the number of beams was optimised the values were 6, 7 and 8% for 7, 5 and 3 IMRT fields, respectively (all $p < 0.001$ compared to 3DCRT). The rectal and bladder volumes irradiated with doses >45 Gy were also reduced by IMRT. The reduction in critical pelvic organ irradiation seen with IMRT may reduce side effects in patients, and allow modest dose escalation within acceptable complication rates. This is now being assessed in a prospective, phase I, dose escalation trial. The advantages of IMRT were maintained with 3–5 IMRT field plans that potentially allow less complex delivery techniques and shorter delivery times. The second application of IMRT under investigation is the use of a concurrent boost, to escalate the dose to the gross tumour volume within the prostate, without increasing rectal and bladder irradiation.

## Conclusions

3DCRT and IMRT represent important technical advances in the delivery of radiation therapy for prostate cancer. Non-randomised studies have demonstrated potential reductions in side effects and improvements in tumour control with dose escalation. The phase III randomised trial at the RMH/ICR has confirmed a significantly reduced incidence of late rectal side effects and currently dose escalation is being tested in trials in Europe and North America. These trials will define the balance between improved local control and late radiotherapy side effects and lead to overall improvements in outcome. Technical challenges for the future include developing and refining of planning systems taking into account organ and patient movement, and the definition of the role of more complex IMRT methods.

## References


1   Dearnaley DP, Khoo VS, Norman A, et al: Comparison of radiation side effects of conformal and conventional radiotherapy in prostate cancer: A randomised trial. Lancet 1999;352:267–272.
2   Dearnaley DP, Nahum A, Lee M, Tait D, Wynne C, Horwich A: Radiotherapy of prostate cancer: Reducing the treated volume. Conformal therapy, hormone cytoreduction and protons. Br J Cancer 1994;70(suppl 22):16.
3   Nutting C, Khoo VS, Walker V, et al: A randomised trial of the use of a customised immobilisation system in the treatment of prostate cancer with conformal radiotherapy. Radiother Oncol 2000;57:1–9.
4   Nutting C, Convery DJ, Cosgrove VP, et al: Reduction of small and large bowel irradiation using an optimised intensity-modulated pelvic radiotherapy technique (IMRT) in patients with prostate cancer. Int J Radiat Oncol Biol Phys 2000;48:649–656.
5   RT01: A randomised trial of high dose therapy in localised cancer of the prostate using conformal radiotherapy techniques (1997). An MRC Radiotherapy Working Party Protocol (available at MRC).



Dr. C. Nutting, FRCR, Academic Department of Radiotherapy and Oncology,
3rd floor Orchard House, Royal Marsden NHS Trust, Downs Road,
Sutton, Surrey SM2 5PT (UK)
Tel. +44 208 661 3271, Fax +44 208 643 8809, E-Mail chrisnutting@cs.com